LITHIUM USE IN BATTERIES

DEMAND AND SUPPLY CONSIDERATIONS

MATERIALS SCIENCE AND TECHNOLOGIES

Additional books in this series can be found on Nova's website
under the Series tab.

Additional e-books in this series can be found on Nova's website
under the e-book tab.

ENERGY SCIENCE, ENGINEERING AND TECHNOLOGY

Additional books in this series can be found on Nova's website
under the Series tab.

Additional e-books in this series can be found on Nova's website
under the e-book tab.

MATERIALS SCIENCE AND TECHNOLOGIES

LITHIUM USE IN BATTERIES

DEMAND AND SUPPLY CONSIDERATIONS

DONALD R. TAYLOR

AND

RYAN I. YOUNG

EDITORS

Nova Science Publishers, Inc.

New York

For permission to use material from this book please contact us:
Telephone 631-231-7269; Fax 631-231-8175
Web Site: http://www.novapublishers.com

NOTICE TO THE READER

The Publisher has taken reasonable care in the preparation of this book, but makes no expressed or implied warranty of any kind and assumes no responsibility for any errors or omissions. No liability is assumed for incidental or consequential damages in connection with or arising out of information contained in this book. The Publisher shall not be liable for any special, consequential, or exemplary damages resulting, in whole or in part, from the readers' use of, or reliance upon, this material. Any parts of this book based on government reports are so indicated and copyright is claimed for those parts to the extent applicable to compilations of such works.

Independent verification should be sought for any data, advice or recommendations contained in this book. In addition, no responsibility is assumed by the publisher for any injury and/or damage to persons or property arising from any methods, products, instructions, ideas or otherwise contained in this publication.

This publication is designed to provide accurate and authoritative information with regard to the subject matter covered herein. It is sold with the clear understanding that the Publisher is not engaged in rendering legal or any other professional services. If legal or any other expert assistance is required, the services of a competent person should be sought. FROM A DECLARATION OF PARTICIPANTS JOINTLY ADOPTED BY A COMMITTEE OF THE AMERICAN BAR ASSOCIATION AND A COMMITTEE OF PUBLISHERS.

Additional color graphics may be available in the e-book version of this book.

Library of Congress Cataloging-in-Publication Data

Lithium use in batteries : demand and supply considerations / editors, Donald R. Taylor and Ryan I. Young.
p. cm.
Includes bibliographical references and index.
ISBN 978-1-62257-037-9 (soft cover)
1. Lithium cells. 2. Lithium industry. I. Taylor, Donald R., 1972- II. Young, Ryan I., 1970-
TK2945.L58L63 2011
669'.725--dc23
2012018739

Published by Nova Science Publishers, Inc. † New York

CONTENTS

PREFACE

Lithium has a number of uses but one of the most valuable is a component of high energy-density rechargeable lithium-ion batteries. Because of concerns over carbon dioxide footprint and increasing hydrocarbon fuel cost (reduced supply), lithium may become even more important in large batteries for powering all-electric and hybrid vehicles. Estimates of future lithium demand vary, based on numerous variables. Some of those variables include the potential for recycling, widespread public acceptance of electric vehicles, or the possibility of incentives for converting to lithium-ion-powered engines. This book addresses some of the issues raised by the increased focus on lithium, including the context of the lithium market into which future lithium-based large batteries must fit, the projected effect of electric and hybrid cars on lithium demand, various estimates for future lithium demand, and obstacles to reaching the more optimistic estimates.

Chapter 1- Lithium has a number of uses but one of the most valuable is as a component of high energy-density rechargeable lithium-ion batteries. Because of concerns over carbon dioxide footprint and increasing hydrocarbon fuel cost (reduced supply), lithium may become even more important in large batteries for powering all-electric and hybrid vehicles.

It would take 1.4 to 3.0 kilograms of lithium equivalent (7.5 to 16.0 kilograms of lithium carbonate) to support a 40-mile trip in an electric vehicle before requiring recharge.

This could create a large demand for lithium. Estimates of future lithium demand vary, based on numerous variables. Some of those variables include the potential for recycling, widespread public acceptance of electric vehicles, or the possibility of incentives for converting to lithiumion-powered engines. Increased electric usage could cause electricity prices to increase. Because of reduced demand, hydrocarbon fuel prices would likely decrease, making hydrocarbon fuel more desirable.

In 2009, 13 percent of worldwide lithium reserves, expressed in terms of contained lithium, were reported to be within hard rock mineral deposits, and 87 percent, within brine deposits. Most of the lithium recovered from brine came from Chile, with smaller amounts from China, Argentina, and the United States. Chile also has lithium mineral reserves, as does Australia.

Another source of lithium is from recycled batteries. When lithium-ion batteries begin to power vehicles, it is expected that battery recycling rates will increase because vehicle battery recycling systems can be used to produce new lithium-ion batteries.

Chapter 2- Use of vehicles with electric drive, which could reduce people's oil dependence, will depend on lithium-ion batteries. But is there enough lithium? Will the authors need to import it from a new cartel? Are there other materials with supply constraints? The authors project the maximum demand for lithium and other materials if electric-drive vehicles expanded their market share rapidly, estimating material demand per vehicle for four battery chemistries. Total demand for the United States is based on market shares from an Argonne scenario that reflects high demand for electric-drive vehicles, and total demand for the rest of the world is based on a similar International Energy Agency scenario. Total material demand is then compared to estimates of production and reserves, and the quantity that could be recovered by recycling, to evaluate the adequacy of supply. The authors also identify producing countries to examine potential dependencies on unstable regions or future cartels.

Chapter 3- The transition to plug-in hybrid vehicles and possibly pure battery electric vehicles will depend on the successful development of lithium-ion batteries. But, in addition to issues that affect performance and safety, there could be issues associated with materials. Many cathode materials are possible, with trade-offs among cost, safety, and performance. Oxides of cobalt, nickel, manganese, and aluminum in various combinations could be used, as could iron phosphates.

The anode material of choice has been graphite, but titanates may be used in the future. Similarly, different materials could be used for other parts of the cell. The authors consider four likely battery chemistries and estimate the quantities of all of these materials that could be required if vehicles with large batteries made significant market inroads, and the authors compare these quantities to world production and resources to identify possible constraints. The authors identify principal producing countries to identify potential dependencies on unstable regions or cartel behavior by key producers. The authors also estimate the quantities of the materials that could be recovered by recycling to alleviate virgin material supply restrictions and associated price increases.

Chapter 4- Following a 2009 investment of $32.9 billion in renewable energy and energy efficiency research through the American Recovery and Reinvestment Act, President Obama in his January 2011 State of the Union address promised deployment of one million electric vehicles by 2015 and 80% clean energy by 2035.

The United States seems poised to usher in its bright energy future, but in reality, industry supply chains still rely on foreign sources for many key feedstock materials. In particular, 43% of the lithium consumed domestically is imported, primarily from Chile, Argentina and China, and in 2010, only one company produced lithium from U.S. resources.

Geothermal brines of the Imperial Valley resources of Southern California have been shown to be especially enriched in lithium but today remain an untapped resource.

Producing lithium battery feedstocks at geothermal production facilities could not only provide the U.S. with much-needed lithium products and by-products, but could provide millions of dollars in added revenue to geothermal developers.

By providing lithium reserve estimates, Imperial Valley production potential and forecasts of the future of the electric vehicles industry, this study aims to relate the imperative of domestic lithium production to the vast potential of U.S. geothermal resources and showcase the benefits of industry adoption of lithium co-production at viable geothermal power plants.

In Lithium Use in Batteries: ISBN: 978-1- 62257-037-9
Editors: D. R. Taylor and R. I. Young   © 2012 Nova Science Publishers, Inc

Chapter 1

LITHIUM USE IN BATTERIES[*]

The United States Department of the Interior

CONVERSION FACTORS AND DATUM

Multiply	By	To obtain
	Length	
meter (m)	3.281	foot (ft)
kilometer (km)	0.6214	mile (mi)
	Mass	
kilogram (kg)	2.205	pound avoirdupois (lb)
metric ton (t)	1.102	ton, short (2,000 lb)
	Energy	
kilowatthour (W)	3,600,000	joule (J)

---

[*] This is an edited, reformatted and augmented version of a U.S. Department of Health and Human Services circular 1371 publication.

ABBREVIATIONS AND ACRONYMS

DOE	U.S. Department of Energy
EVq	electric vehicle
HEV	hybrid electric vehicle
Li	lithium
Ni-MH	nickel-metal hydride
PHEV	plug-in hybrid electric vehcle
RBRC	Rechargeable Battery Recycling Corporation
INMETCO	International Metals Reclamation Company, Inc.
USGS	U.S. Geological Survey

ABSTRACT

Lithium has a number of uses but one of the most valuable is as a component of high energy-density rechargeable lithium-ion batteries. Because of concerns over carbon dioxide footprint and increasing hydrocarbon fuel cost (reduced supply), lithium may become even more important in large batteries for powering all-electric and hybrid vehicles.

It would take 1.4 to 3.0 kilograms of lithium equivalent (7.5 to 16.0 kilograms of lithium carbonate) to support a 40-mile trip in an electric vehicle before requiring recharge.

This could create a large demand for lithium. Estimates of future lithium demand vary, based on numerous variables. Some of those variables include the potential for recycling, widespread public acceptance of electric vehicles, or the possibility of incentives for converting to lithiumion-powered engines. Increased electric usage could cause electricity prices to increase. Because of reduced demand, hydrocarbon fuel prices would likely decrease, making hydrocarbon fuel more desirable.

In 2009, 13 percent of worldwide lithium reserves, expressed in terms of contained lithium, were reported to be within hard rock mineral deposits, and 87 percent, within brine deposits. Most of the lithium recovered from brine came from Chile, with smaller amounts from China, Argentina, and the United States. Chile also has lithium mineral reserves, as does Australia.

Another source of lithium is from recycled batteries. When lithium-ion batteries begin to power vehicles, it is expected that battery recycling rates will increase because vehicle battery recycling systems can be used to produce new lithium-ion batteries.

INTRODUCTION

Lithium is the lightest metal and the least dense solid element and, in the latter part of the 20th century, became important as an anode material in lithium batteries. The element's high electrochemical potential makes it a valuable component of high energy-density rechargeable lithium-ion batteries. Other battery metals include cobalt, manganese, nickel, and phosphorus. Batteries are ubiquitous in advanced economies, powering vehicle operations, sensors, computers, electronic and medical devices, and for electrical grid-system load-leveling and are produced and discarded by the billions each year. There is concern that the demand for battery metals could increase, possibly to the point at which a shortage of these metals will occur. Lithium is of particular interest because it is the least likely of the battery metals to be replaced by substitution because it has the highest charge-toweight ratio, which is desired for batteries in transportation applications.

Lithium batteries already enjoy a sizeable market, powering laptop computers, cordless heavy-duty power tools, and hand-held electronic devices. But an even greater market could exist for lithium as a component of electric and hybrid vehicle batteries and for alternative energy production.

Concerns about the carbon dioxide footprint of hydrocarbon-based powerplants and internal-combustion-powered automobiles, the projected hydrocarbon shortage (which could mean high prices) in coming years, and U.S. dependency on foreign hydrocarbon fuels have spurred great interest in alternative energy sources.

Electric-powered vehicles are expected to take market share from internal-combustion-powered vehicles in the future. Large batteries are and will continue to be needed for powering all-electric and hybrid vehicles and also for load leveling within solar- and wind-powered electric generation systems. Research on lithium for use in large batteries is in advanced stages. Future light vehicles will potentially be powered by electric motors with large, lightweight batteries, and lithium is a particularly desirable metal for use in these batteries because of its high charge-to-weight ratio. Table 1 shows the plans of automobile manufacturing companies, as of 2010, for introducing lithium-ion-powered vehicles.

This report addresses some of the issues raised by the increased focus on lithium, including the context of the lithium market into which future lithium-based large batteries must fit, the projected effect of electric and hybrid cars on lithium demand, various estimates for future lithium demand, and obstacles to reaching the more optimistic estimates.

Table 1. Announced introductions of lithium-ion powered automobiles through July 2010

Automobile manufacturer	Vehicle name (type)	Date of introduction	Comments
Audi	E-Tron (pure electric)	2013	Concept sports car.
			Lithium-ion battery powered motor on each wheel.
BYD (China)	E6 (pure electric)	2010	Currently being tested by Shenzhen Taxi Co.
			Iron-based lithium-ion battery.
			About $43,000 retail (before 20 percent government subsidy).
BMW	Mega City (pure electric)	2013	Planning stage.
Chrysler	Fiat 500EV (pure electric)	2012	Lithium-ion battery pack.
			Estimated range 80-100 miles.
			Expect to use U.S.-produced battery.
Ford	Ford Fusion BEV (pure electric)	2011	Currently testing concept cars.
			Lithium-ion battery pack.
			Capacity of 23 kWh and a range of up to 75 miles.
			Charging the batteries will take between 6 and 8 hours, using a household 230-V electricity supply.
General Motors	Chevrolet Volt (pure electric)	2011	Concept car exists.
			Powered by lithium-ion battery pack, which will be manufactured in the United States.
Honda	FCX Clarity (fuel cell)	2010	Hydrogen-powered fuel cell.
			Lithium-ion battery for supplemental power.
Hyundai	Blue-Will (plug-in hybrid)	2012	Lithium-ion battery powered.
Mercedes Benz	SLS AMG (pure electric)	2013	Concept sports car.
			Hydrogen fuel cell plus lithium-ion battery.
Nissan	LEAF (plug-in hybrid)	2012	May 26, 2010, broke ground for:
			Auto plant 150,000-vehicle-per-year capacity.
			Lithium-ion battery plant 200,000 unit-per-year capacity.
Tesla	Roadster (pure electric)	2008	Currently marketing electric automobiles.

Automobile manufacturer	Vehicle name (type)	Date of introduction	Comments
			Lithium-ion battery pack (liquid cooled); 900 pounds, storing
			56 kWh of electric energy, delivering 215 kW of electric power
Toshiba-Mitsubishi JV	Unspecified	unspecified	Hopes to sell lithium-ion batteries for future Mitsubishi Motors vehicles.
Toyota	Prius-PHV (plug-in hybrid)	2010	Test program, 500 vehicles placed worldwide.
			First generation lithium-ion battery.
			Maximum range (fully electric) = 13 miles.
			Maximum speed (fully electric) = 60 mph.
Volkswagen	e-Golf (pure electric)	2013	To be tested in 2011.
			Air-cooled 26.5 kW lithium-ion battery pack.
			Expect 93 miles on one charge.

[Data are from Ford Motor Company (2009), Kanellos (2009), Toyota Motor Sales, USA., Inc. (2009), Abuelsamid (2010), American Honda Motor Co., Inc. (2010), China Car Times (2010), Ewing (2010), General Motors Company (2010), Green Car Reports (2010), Murray (2010), Nissan (2010), Osawa and Takahashi (2010), and Tesla Motors (2010). JV, joint venture; kW, kilowatt; kWh, kilowatthour; mph, miles per hour; V, volt].

LITHIUM DEMAND

Lithium Consumption Statistics

Apparent consumption of lithium in the United States has been recorded since at least 1900 (fig. 1) and includes only imports minus exports because lithium is not mined domestically. Significant apparent consumption began in the 1950s, peaked in 1974, and has shown a slightly decreasing trend since 1974. The consumption figures do not include lithium contained in imported finished assemblies, for example, lithium contained in batteries (almost all of which are manufactured overseas) that are within computers, electronic devices, and tools. n 2007 and 2008, an estimated 25,400 metric tons (t) of lithium was used each year for various products worldwide. Owing to the general downturn in the world economies, total lithium use in 2009 decreased to approximately 18,000 t. Table 2 lists the percentage of lithium used worldwide in each product during those 3 years, as estimated by the U.S. Geological Survey (Jaskula, 2008–2010). Of particular significance, the lithium use in batteries decreased by approximately 2,062 t, or 35 percent, between 2008 and 2009. Lithium use in rechargeable batteries increased from zero in 1991 to 80 percent of the market share in 2007, with 1992 being the first time nickel-cadmium and nickel-metal-hydride (NiMH) batteries started to be replaced by lithium-ion batteries (fig. 2). The greater charge-to-density (power-toweight) ratio of lithium is favorable for electronic devices and has helped to drive this trend.

Table 2. World market shares for various lithium end-uses from 2007 through 2009

End-use	2007	2008	2009
World market share:			
Ceramics and glass	18%	31%	30%
Batteries	25%	23%	21%
Lubricating greases	12%	10%	10%
Pharmaceuticals and polymers	7%	7%	7%
Air conditioning	6%	5%	5%
Primary aluminum (alloying)	4%	3%	3%
Other	28%	21%	24%
World production, in metric tons of contained lithium	25,400	25,400	18,000

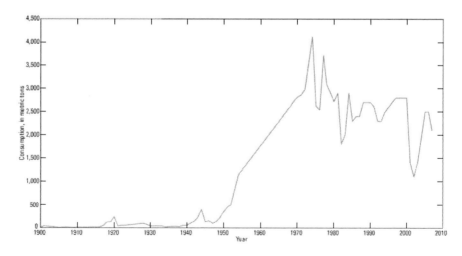

Figure 1. Chart showing consumption of lithium in the United States from 1900 through 2007. Values are in metric tons. Data are from U.S. Geological Survey (2010).

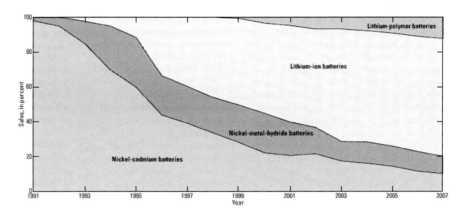

Figure 2. Chart showing sales of rechargeable batteries worldwide from 1991 through 2007. Values are expressed as percentage of total global sales of rechargeable batteries. Data are from Wilburn (2007) and Takashita (2008).

Effect of Electric and Hybrid Cars on Lithium Demand

Although electric vehicles have existed for more than a century, Toyota's hybrid Prius was the first to have commercial success. Now many automobile manufacturers are expanding into cutting-edge electromotive powertrains (table 1; Hsiao and Richter, 2008). Electric cars are characterized as—all electric (EV), hybrid (HEV), or plug-in hybrid (PHEV) vehicles. Concerns about the dependence on imports of oil and about the carbon footprint of internal-combustion engines in current automobile industry products have created this interest in electric vehicles. In fact, the U.S. Government planned to provide $11 billion in loans and grants to car and battery makers to reduce the country's dependence on foreign oil (Smith and Craze, 2009). These funds will be targeted for research and development and for production and recycling facilities.

Through 2010, the predominant battery technology powering experimental electric vehicles has been NiMH, although the General Motors EV–1 was powered by a lead-acid battery. NiMH batteries offer proven performance, reasonable energy density, and thermal stability. They are also large, heavy, and expensive and require a long time to charge compared with lithium-ion batteries. In 2008, attention was directed toward lithium-ion batteries as an alternative, although safety, longevity, and cost were of concern (Hsiao and Richter, 2008). The high charge-to-weight ratio of lithium makes the lithium-ion battery much lighter than the NiMH battery, which is desirable for powering electric vehicles. Although NiMH batteries are affected by the "memory effect" (the battery loses its capacity when it is recharged without being fully depleted), lithium-ion batteries are not (PlanetWatch, 2009). These qualities have helped to bring lithium-ion technology to the forefront as the object of extensive research (Gaines and Cuenca, 2000). In automotive applications, individual cells are typically connected together in various configurations and packaged with associated control and safety circuitry to form a battery module (Anderson, 2009). Therefore, though most research is directed toward improving lithium-ion battery technology at the cell level, research is likely to also be directed toward determining the most effective cell configurations and packaging.

Depending on lithium-ion battery chemistry, it would take 1.4 to 3.0 kilograms (kg) of lithium equivalent (7.5–16.0 kg of lithium carbonate) to support a 40-mile trip in an electric vehicle before requiring recharge (Gaines and Nelson, 2009). If the trend toward replacing internal combustion engine vehicles with electric vehicles continues and lithium-ion batteries become the preferred power source for electric vehicles, then a large demand for lithium carbonate could potentially be generated.

Estimates of Future Lithium Demand

Several authors have estimated future lithium demand using certain assumptions and projections of electric car demand. Gaines and Nelson (2009) optimistically calculate that U.S. annual demand for electric vehicles might require as much as 22,000 t of lithium (117,000 t of lithium carbonate) by 2030, and as much as 54,000 t (287,000 t of lithium carbonate) by 2050, assuming the lithium-nickel-cobaltgraphite chemistry that is currently popular. This projection further assumes continued growth in all automobile sales, 52 percent electric vehicle penetration in 2030, and 90 percent in 2050, which Gaines and Nelson admit are optimistic assumptions.

Tahil (2007, 2008) expressed concern that, if the 60 million cars that are produced worldwide each year were totally replaced with plug-in hybrids, each having a 5-kilowatt battery (requiring about 1.40 kg of lithium carbonate), demand for lithium carbonate would be 420,000 t annually, which is nearly 5 times the current lithium carbonate production. This would place an unsustainable demand on lithium resources because of geochemical constraints in extracting the product from known deposits. In July 2009, Chemetall GMBH, a division of Rockwood Holdings, Inc., which holds 30 percent of the global lithium carbonate market share, estimated that lithium carbonate demand in 2020 would be either 145,000 t (42 percent automotive) or 116,000 t (27 percent automotive), depending on if Gaines and Nelson's (2009) or Tahil's (2007, 2008) scenario was used (Haber, 2008; Chemetall, 2009).

These new lithium demand estimates, which are derived from expected use of lithium in next-generation electric vehicles, vary. One must understand the assumptions, including the potential for recycling, that underlie published estimates. Before any of the more optimistic estimates for lithium demand are actualized, some significant obstacles must be overcome. These are summarized below.

The lithium-based battery packs used in automobiles are much larger than the small lithium-ion batteries currently being produced for use in electronic devices. While technical testing has been encouraging, large-scale lithium-ion battery packs have not been fully market tested (table 1). The level of use that electric vehicles achieve will depend in part on consumer acceptance. Product safety, convenience of use, reliability, and cost of purchase and operation are likely to influence consumer acceptance.

Electric-powered vehicles currently cost more than equivalently-sized vehicles powered by internal combustion. For electric vehicles ti become cost effective, the savings from using electric power would have to offset the incremental capital cost (Simpson, 2006) and the cost of operating the vehicle.

Competition between the price of electricity and the price of grasoline will affect the adoption of electric vehicles. The price of gasoline is set by market forces and changes as levels of consumption change. The price for electric power is usually set by regulatory bodies and is therefore less responsive to changes in use.

The adoption of electric vehicles is likely to be constrained by the capacity of the electricity grid unless electric vehicles are recharged during off-peak times. Changes in the pricing of recidential electricity and the use of devices such as smart meters would have an affect (Xcel Energy Inc., 2010).

LITHIUM SUPPLY

The two most important sources of lithium are a hard silicate mineral called spodumene, which is found in pegmatites, and brine lake deposits that contain lithium chloride. In 2009, of the worldwide reported lithium reserves, expressed in terms of contained lithium, 13 percent was reported to be within hard rock mineral deposits, and 87 percent, within brine deposits (Jaskula,

2009, 2010). Reserves are known quantities that are presently economic to exploit (U.S. Bureau of Mines and U.S. Geological Survey, 1980).

Production of lithium carbonate from spodumene is more energy intensive compared with production from brine and is more costly because of added extraction and beneficiation challenges. Comparing lithium production from these two sources between 1990 and 2008 (fig. 3), the compound annual growth rate (CAGR) of lithium from brine deposits has been 11.7 percent per year, whereas lithium's CAGR from hard rock deposits has been 7.4 percent per year. Overall, the CAGR of lithium production from 1990 to 2008 was 10.4 percent.

In 2008, Australia produced most of the lithium from hard rock (table 3). Most of the lithium recovered from brine came from Chile, with smaller amounts from China, Argentina, and the United States. Chile has two producers of lithium products, Sociedad Química y Minera de Chile S.A. (SQM) and Chemetall SCL. Both companies operate at the Salar de Atacama, Chile, and account for more than 65 percent of the world lithium market (Lithium Site, 2009).

The Salar de Atacama holds about 29 percent of the world's known lithium resources, and together, the salt lakes of South America (Argentina, Bolivia, and Chile) contain about 75 percent of the world's known lithium resources (Jaskula, 2010, p. 93). At SQM's operation, the brine is pumped from about 40 meters (m) below the surface and then placed in surface ponds, where it is exposed to evaporation under conditions of high heat, low humidity, and strong surface winds (Energy Investment Strategies, 2008). The resulting lithium chloride concentrate is further treated with sodium carbonate to produce the desired lithium carbonate. In 2008, SQM's annual capacity of lithium carbonate production at the Salar de Atacama was expanded to 40,000 metric tons per year (t/yr); meanwhile, Chemetall maintained capacity of 27,000 t/yr at the Salar de Atacama (Chemetall, 2009; de Solminihac, 2009).

Argentina has at least two brine deposits of importance. The Salar del Hombre Muerto operation, which is at 3,962 m above sea level and operated by FMC Corporation, is recovering lithium using a proprietary separation process (Lithium Site, 2009). Production capacity at the Salar del Hombre Muerto, is 12,000 t/yr of lithium carbonate and 6,000 t/yr of lithium chloride (Tahil, 2007). In January 2007, the brine operation Salar del Rincon opened pilot-plant-scale operations in Argentina (Tahil, 2007); the operations were still under development in 2010.

China is also a major lithium producer (13 percent of world production of contained lithium).

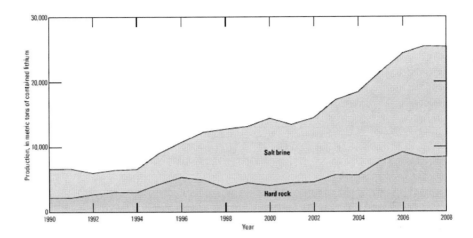

Figure 3. Chart showing production of lithium, by deposit type, worldwide from 1990 through 2008. Values are in metric tons of contained lithium. Data are from U.S. Bureau of Mines (1992–1995) and U.S. Geological Survey (1996–2009).

Salt lakes are widely distributed across China's western Qinghai, Tibet, Xinjaing and inner Mongolia, with rich resources of boron, lithium, magnesium, and potassium (Ma, 2000). China is currently developing three brine lake deposits—the Taijinaier salt lake in Qaidam Basin, Qinghai Province, north of Tibet; the Dangxiongcuo (DXC) salt lake in southwestern Tibet; and the Zhabuye salt lake in western Tibet (Tahil, 2007, p. 10). With the success of a 500-t/yr pilot plant at Taijinaier salt lake, CITIC Guoan Scientific and Technical Company inaugurated a 35,000-t/yr lithium carbonate plant in 2007 (Zhang, 2009). The Canadian company Sterling Group Ventures is considering development of a 5,000-t/yr lithium carbonate plant at DXC salt lake (Zhang, 2009). The Zhabuye salt lake— the third largest salt lake (in terms of area) in the world—is at 4,400 m above sea level and is the largest lithium deposit in China (Green Energy News, 2008). In 2008, Baiyin Zhabuye Lithium Industries Co., Ltd, produced 2,000 t of lithium carbonate and lithium hydroxide from this deposit and has government approval to increase lithium carbonate production capacity by 12,000 t/yr (Zhang, 2009).

The brines of the Salar de Uyuni in Bolivia are also a potential source of lithium carbonate. The deposit contains approximately 9 million metric tons of lithium and could account for as much as 50 percent of the global lithium reserves; it is currently under consideration for development (Tahil, 2007).

The government of Bolivia has sought to keep its development under government auspices and has begun to build a 30,000-t/yr lithium carbonate production facility at the deposit (New Tang Dynasty Television, 2009).

Table 3. World production of lithium from minerals and brine in 2008, by country

Country[1]	Deposit type	Lithium product	Production
Production from minerals:			
Australia	Spodumene	Concentrate	6,280
Brazil	Various	Concentrate	160
Canada[2]	Spodumene	Concentrate	690
China	Various	Li_2CO_3	880
Portugal	Lepidolite	Concentrate	700
Zimbabwe	Various	Concentrate	500
Total			9,210
Production from brine:			
Argentina[3]	NA	Li_2CO_3	1,880
	NA	LiCl	1,290
Chile[3]	NA	$Li_2 CO_3$	9,870
	NA	LiCl	720
China	NA	Li_2CO_3	2,410
United States[4]	NA	$Li_2 CO_3$	1,710
Total			17,900

[Values are in metric tons of contained lithium. Production data are estimated and rounded to no more than three significant digits. Table includes data available through April 1, 2009. Data are from Jaskula (2008) and Tahil (2008). LiCl, lithium chloride; Li2CO3, lithium carbonate; NA, not available]

[1] Other countries produce small amounts of lithium but are not included here.

[2] Based on all Canada's spodumene concentrates (Tantalum Mining Corp. of Canada Ltd., Tanco property).

[3] New information was available from Argentine and Chilean sources, prompting major revisions in how lithium production was reported.

[4] The estimate for the United States is taken as the suggested production of Chemetall's Clayton Valley mine at Silver Peak, Nevada, as reported by Tahil (2008, p. 20).

These brine lake deposits and other deposits not specifically discussed in this report each have unique characteristics with respect to lithium content, salt chemistry, and general ease (cost) of processing. The market together with the governments' willingness to subsidize lithium supply and demand, either directly or indirectly through tax-modified behavior, will determine where the lithium is produced and the reserve estimates of the moment.

Brine-originated lithium carbonate, the primary ingredient in lithium-ion batteries, accounted for about 67 percent of lithium production (excluding the United States) in 2008. Lithium carbonate can be made from lithium concentrate, but it is more expensive to do so (Tahil, 2007). Supply and demand for lithium is currently balanced. Expansion of worldwide brine operations is dependent upon lithium carbonate from brines beings less expensive than from competing sources and an expanding lithium-based battery market to serve an assumed growing electric vehicle market.

Lithium Carbonate Prices

The prices of lithium carbonate (Li_2CO_3) imported into the United States from 1989 through 2008 are shown in figure 4. The unit value of U.S. imports was used because it is presumably more representative of world prices than the unit value of U.S. exports, which are more refined and a higher priced form of lithium carbonate.

The unit value of imports of lithium carbonate into the United States decreased from 1995 through 1999, reflecting growth in supply of lithium carbonate from low-cost brine deposits (fig. 3). The period from 1999 through 2005 experienced nondynamic supply and demand activity. From 2006 through 2008, increased demand for lithium carbonate resulted in higher prices, leading to increased investment in exploration and new capacity development.

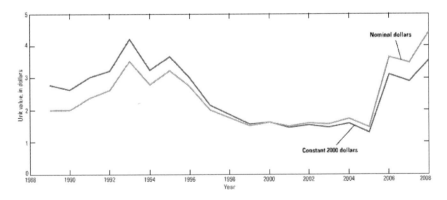

Figure 4. Chart showing the unit value of imports of lithium carbonate into the United States from 1989 through 2008. Values are in dollars per kilogram of lithium carbonate. Data are from U.S. Geological Survey (1996–2009) and U.S. Bureau of Mines (1992–1995).

Because automobile batteries are expected to become (although they are not yet) a major factor in total battery demand, there is some concern about whether world lithium reserves will be sufficient to supply a future surge in automobile-generated lithium demand (Tahil, 2007, 2008). Others are less concerned (Pease, 2008; Beckdorf and Tilton, 2009; Gaines, 2009; Gaines and Nelson, 2009). Historically, reported reserve levels were not limits but rather were indicative of actual market conditions at the time of assessment.

LITHIUM BATTERIES

Battery Types

Lithium batteries contain metallic lithium and are not rechargeable. The button-sized cells that power watches, hand-held calculators, and small medical devices are usually lithium batteries. These are also called primary lithium batteries, and they provide more useable power per unit weight than do lithium-ion batteries (called secondary batteries). Lithium-ion batteries use lithium compounds, which are much more stable (less likely to oxidize spontaneously) than the elemental lithium used in lithium batteries (Green Batteries, 2009).

There are many lithium-ion battery types and configurations. These batteries are not generally available in standard household sizes but rather are manufactured specifically for a particular electronic device. It is possible to classify lithium-ion battery types according to battery chemistry and packaging. Table 4 lists the most common rechargeable lithium-ion chemistries.

One or more of the lithium-ion battery chemistries displayed in table 4 or another entirely different lithium-ion battery chemistry may become the basis for the future electric vehicle power supply. The major difference between batteries for electronics and batteries for electric vehicles will be size. Increased size can be obtained by making assemblies of small cells or by developing singular large cells.

A detailed cost and technical study of lithium-ion battery development for automobiles is beyond the scope of this report but can be found in Gaines and Cuenca (2000), Hsiao and Richter (2008), Anderson and Patiño-Echeverri (2009), Burke and Miller (2009), Nelson, Santini, and Barnes (2009), and National Institute of Advanced Industrial Science and Technology [Japan] (2009).

Table 4. Common lithium-ion rechargeable battery chemistries

Cathode name and chemistry	Cell voltage		Electric charge, Ah/g		Energy density, Wh/kg	Applications
	Maximum	Nominal	Anode	Cathode		
Cobalt, Li(Ni 0.85, Co 0.1, Al 0.05)O$_2$	4.2	3.6	0.36	0.18	100–150	Cell phone, cameras, laptops.
Manganese (spinel), (LiMn$_2$)O$_4$	4.0	3.6	0.36	0.11	100–120	Power tools, medical equipment.
Nickel, cobalt, manganese, Li(Ni 0.37, Co 0.37, Mn 0.36)O$_2$	4.2	3.6	0.36	0.18	100–170	Power tools, medical equipment.
Phosphate, (Li,Fe)PO$_4$	3.65	3.25	0.36	0.16	90–115	Power tools, medical equipment.

[Associated data are in specified units. Data are from Buchmann (2006), Burke and Miller (2009), and Gaines and Nelson (2009). Ah/g, ampere-hours per gram; Al, aluminum; Co, cobalt; Fe, iron; Li, lithium; Ni, nickel; Mn, manganese; O, oxygen; PO4, phosphate; Wh/kg, watthours per kilogram].

Rechargeable lithium-ion batteries can be categorized by packaging in the following categories:

- cylindrical cells, which are the most widely used packaging for wireless communication, mobile computing, biomedical instruments, and power tools (Buchmann, 2004)
- prismatic cells, which were developed in the early 1990s, are made in various sizes and capacities, and are custom made for electronic devices, such as cell phones (Buchmann, 2004)
- pouch cells, which were introduced in 1995, permit tailoring to the exact dimensions of the electronic device manufacturer, and are also easily assembled into battery packs as needed (Buchmann, 2004)

Battery Production

Through 2009, lithium-ion (rechargeable) battery production in the United States has been limited to small-scale, high-profit-margin niche markets, such as medical, military, or space applications, and the greater part of general-use lithium-ion batteries has been produced in China, Japan, and the Republic of Korea (Wilburn, 2008, p. 3).

In 2009, General Motors announced the construction of a lithium-ion battery pack production plant to be located in Warren, Michigan, which will produce vehicle batteries for the its new electric car, the Volt, which is scheduled to premier in 2011 (Brooke, 2009).

Japan is a major producer of lithium-based batteries. In 2009, lithium-based batteries accounted for 43 percent of the total volume (4.34 billion units) of batteries produced in Japan—47 percent of lithium batteries were primary lithium batteries, and 53 percent were lithium-ion batteries (Battery Association of Japan, 2010).

Since lithium batteries first entered the market in 1993, about 45,000 t of lithium has been incorporated into these batteries worldwide. Figure 5 shows the annual and cumulative lithium battery production from 1993 through 2008.

Between 1993 and 2008, the lithium battery market consisted of nonrechargeable (primary) and rechargeable (secondary) batteries for electronic devices.

Only about 0.2 percent of lithium produced went to automobile batteries in 2008 (Wilburn, 2008, p. 13).

Battery Recycling

In 2009, an estimated 3,700 t of lithium, contained in scrap batteries, became available to the world market. This estimate was determined by applying a Gaussian distribution to the annual production of batteries (expressed as contained lithium) for 2000–2008 and factoring in the average life of a lithium battery [assumed to be 4 years based on Dan's Data (2008), and Mah (2007)]. The actual amount of lithium recovered (worldwide) from recycled batteries in 2009 is not available for comparison to the amount available for recovery.

In the United States, it is unlikely that more than 20 percent of the batteries available for recycling actually were recycled. Europe, however, has stronger battery collection laws. Most scrap batteries in the United States have likely been sequestered either in homes and businesses or released to municipal solid waste to be retired to landfills or combusted. In 2006, lithium-ion batteries were not considered to be a hazardous waste in the United States (Mitchell, 2006).

When lithium-ion batteries begin to power vehicles, it is expected that battery recycling rates will increase because vehicle battery recycling systems, based on the lead-acid model currently in place, can be used to produce new lithium-ion batteries. Recycling of electric vehicle batteries could provide 50 percent of the lithium requirement for new batteries by 2040 (Chemetall, 2009; Gaines, 2009).

Most if not all types of batteries can be recycled. It costs about $1,100 to $2,200 to recycle 1 t of batteries of any chemistry and size (including small cells), except automobile batteries. Significant subsidies are still required from manufacturers, agencies, and governments to support the battery recycling programs (Buchmann, 2009).

A high-energy (100 ampere-hour) battery processed through recycling would return about 169 kg of lithium carbonate, 38 kg of cobalt, and 201 kg of nickel [calculated from data reported by Hsiao (2008, p. 22)]. This estimate is based on the assumptions that the cost of recycling large automobile batteries is similar to that for small batteries; the automobile battery cathode chemistry will be Li[Ni 0.8, Co 0.15, Al 0.05]O_2, and 98 percent of the metal will be recovered in recycling. At 2008 prices (normalized to 2000 dollar basis) for lithium carbonate and cobalt-nickel metals, the value of the recovered materials would be about $6,400. At 2009 prices, which were very similar to 2005 prices for these materials, the value of the recovered materialswould be about $4,100. Metal pricing will be very important to recycling profitability. Lithium carbonate return contributes only about 10 percent of the total monetary return. If research takes cathode technology to less expensive metals, such as manganese and phosphorus, then the economic attractiveness of recycling these batteries could diminish, perhaps to a point at which recycled lithium carbonate cannot be counted on to supplant pressure on in-ground lithium resources.

The same methodology and assumptions applied to the high-power (10 ampere-hour) battery cut all of the monetary returns by roughly one-half, placing the recycling decision very close to the break-even level. The economics of lithium ion battery recycling needs more research. Xu and others (2008) reviewed the research conducted on this subject through 2007.

The Rechargeable Battery Recycling Corporation (RBRC) was founded in 1994 to promote recycling of rechargeable batteries in North America (table 5). RBRC is a nonprofit organization that collects batteries from consumers and businesses and sends them to North American recycling organizations, such as International Metals Reclamation Company, Inc. (INMETCO) and Toxco Inc. Since 1992, Sony has partnered with Sumitomo Metals to recover cobalt from used lithiumi-ion batteries (Hsiao, 2008).

For most lithium-ion batteries, lithium represents less than 3 percent of the production cost; nickel and cobalt are the biggest economic drivers of recycling (Hamilton, 2009). Toxco is North America's leading battery recycler and has been recycling single-charge and rechargeable batteries used in electronic devices and industrial applications since 1992 at its Canadian facility in Trail, British Columbia (Hamilton, 2009). Toxco can recover up to 98 percent of the lithium carbonate from lithium waste but focuses on cobalt and nickel (Hsiao, 2008).

Table 5. European and North American lithium battery recyclers

Region/country	Company	City, State/Province/region	Capacity
		Europe	
Switzerland	Batrec Industrie AG	Wimmis, Bern	5,000
France	Citron	Rogerville, Seine-Maritime	130,000
	Eurodieuze Industrie	Dieuze, Moselle	NA
	Recupyl	Domène, Isère[1]	110
	S.N.A.M.	Viviez, Aveyron	4,000
Belgium	Umicore	Olen, Antwerp	3,000
		North America	
Canada	Toxco, Inc.	Trail, British Columbia	NA
Canada	Xstrata Nickel International	Falconbridge, Ontario	3,000
United States	Toxco	Lancaster, Ohio	NA

[1] Pilot plant.

LITHIUM BATTERY OUTLOOK

There exists already a large (billions of units per year) market for lithium and lithium-ion batteries, which are used to power hand-held electronic devices and for military purposes. These can be and are recycled using established practices. However, the recycling rate is unknown. Those batteries that are not recycled either go to landfills or remain uncollected at the user level.

If and when the electric motor replaces the internal combustion engine in cars and trucks, the demand for lithium as a major component of batteries, which is the focus of battery research, should increase accordingly. Lithium recycling for the increment of demand represented by automobile batteries that contain lithium should be practical and economical. The recycling would not only recover lithium but would also recover the more expensive metals, including cobalt and nickel. Lithium battery recyclers are already investing in capacity to do just that (Toxco Inc., 2009).

Available data are insufficient to project future lithium demand with certainty. Existing projections are speculative and largely assume a regular progression of automobile sales and an increasing share for electric vehicles. If the general economy remains constrained, then demand for lithium will likely be constrained accordingly. For the electric car share of the automobile market to grow, the relative cost of electric cars will have to decrease so that they are competitively priced compared with internal combustion-powered cars. Also, the cost of electric car batteries would have to drop with economies of scale. This, in turn, would be accompanied by a scale-up of the current (2010) lithium-ion battery technology to batteries of appropriate size for automobiles. To date, this scale-up is indicated, with a heavy research and development focus by battery producers.

Figure 6 shows the sales of hybrid automobiles and crude oil prices from 2000 through 2009. One should not infer a correlation of electric vehicle sales and oil prices from the figure. It is more likely that both are codependent on general economic activity levels. If lithium-containing batteries replace hydrocarbons for powering automobiles, then there will be upward pressure on the prices of the active metals that make up the cathodes of these batteries and possible downward pressure on hydrocarbon prices.

Existing lithium carbonate suppliers believe that they can accommodate the growing demand occasioned by a growing electric car market into the future (Chemetall, 2009; de Solminihac, 2009). The countries with extensive, relatively low-cost, lithium brine deposits are Argentina, Bolivia, Chile, China, and the United States. The marginal cost of lithium carbonate is mostly fixed by the cost of producing lithium carbonate from hard-rock spodumene deposits in Australia.

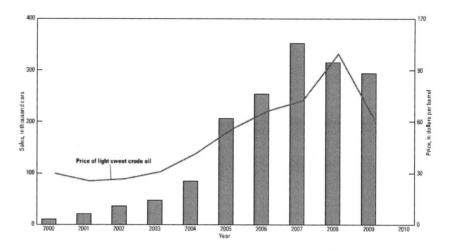

Figure 6. Graph showing sales of hybrid automobiles in the United States and the price of light sweet crude oil from 2000 through 2009. Figures are in numbers of automobiles sold and dollars per barrel. Data are from Hsiao (2008), Hybrid Cars (2010), Truck Trend (2009), and U.S. Department of Energy (2010).

REFERENCES

Abuelsamid, Sam, 2010, Volkswagen releases details of new lithium ion e-Golf, Jetta plug-in coming: Volkswagen news release, May 3, accessed July 7, 2010, at http://www.autoblog.com/2010/05/03/ volkswagen-releases-details-of-new-lithium-ion-e-golf/.

Anderson, D.L., and Patiño-Echeverri, Dalia, 2009, An evaluation of current and future costs for lithium-ion batteries for use in electrified vehicle powertrains: Durham, N.C., Nicholas School of the Environment, Duke University masters project, 44 p., accessed November 17, 2009, at http://dukespace.lib.duke.edu/dspace/bitstream/10161/1007/1/ Li-Ion_ Battery_costs_-_MP_Final.pdf.

Battery Association of Japan, 2010, Total battery production statistics: Battery Association of Japan statistical report, accessed January 12, 2010, at http://www.baj.or.jp/e/ statistics/01.html.

Beckdorf, A.Y., and Tilton, J.E., 2009, Using the cumulative availability curve to assess the threat of mineral depletion— The case of lithium: Golden, Colo., Colorado School of Mines technical paper, 33 p., accessed December 2, 2009, at http://inside.mines.edu/UserFiles/ File/ economics Business/Tilton/The_Case_of_Lithium.pdf.

Brooke, Lindsay, 2009, G.M. to make batteries for Volt in Michigan: New York Times, January 11, 3 accessed January 28, 2010, at http://www. nytimes.com/2009/01/11/ automobiles/autoshow/11BATTERY.html.

Buchmann, Isidor, 2004, Battery packaging—A look at old and new systems: Battery University, accessed January 11, 2010, at http://www.battery university.com/ print-partone-9.htm.

Buchmann, Isidor, 2006, The high-power lithium-ion: Battery University, accessed January 11, 2010, at http://www.batteryuniversity.com/print-partone-9.htm.

Buchmann, Isidor, 2009, Recycling batteries: Battery University, accessed January 11, 2010, at http://www.batteryuniversity.com/print-partone-9. htm.

Burke, Andrew, and Miller, Marshall, 2009, Performance characteristics of lithium-ion batteries of various chemistries for plug-in hybrid vehicles: International Battery, Hybrid, and Fuel Cell Electric Vehicle Symposium, 24, Stavenger, Norway, May 13–16, 2009, presentation, 13 p.

Chemetall, 2009, Lithium applications and availability: Chemetall statement to investors, July, 28 p., accessed January 4, 2010, at http://www. chemetall.com/fileadmin/files_chemetall/Downloads/Chemetall_Li-Supply_ 2009_July.pdf.

China Car Times, 2010, BYD E6 pure electric to launch in August: China Car Times, June 2, accessed July 7, 2010, at http://www. chinacartimes.com /2010/06/02/ byd-e6-pure-electric-to-launch-in-august/.

Dan's Data, 2008, Hatin'on lithium ion: Independent technical review of lithium-ion batteries: Dan's Data, accessed January 13, 2010, at http://www.dansdata.com/gz042.htm.

de Solminihac, P.T., 2009, A statement to investors from Patricio de Solminihac, executive vice-president and chief operating officer: Sociedad Química y Minera de Chile S.A. corporate presentation, September, 21 p., accessed January 4, 2010, at http://www.sqm.com/PDF/Investors/ Presentations/en/CorporatePresentation_Sept09-en.pdf.

Energy Investment Strategies, 2008, A short commercial history of lithium: Energy Investment Strategies, accessed October 6, 2009, at http://www.

energyinvestmentstrategies.com/2008/11/15/ a-short-commercial-history-of-lith.

European Battery Recycling Association, 2009, About EBRA: European Battery Recycling Association, 23 p., accessed January 14, 2010, at http://wwwebrarecycling.org/docs/activities/MEMBERSHIPS/bookletEBR Aabout2009.pdf.

Ewing, Jack, 2010, BMW boasts battery power: Sydney Morning Herald, July 3, accessed July 7, 2010, at http://www.smh.com.au/business/world-business/ bmw-boasts-battery-power-20100702-zu8v.html.

Ford Motor Company, 2009, Ford battery electric vehicles move closer to consumer use: Ford Motor Company, accessed July 7, 2010, at http:// www.thefordstory.com/ green/ford-battery-electric-vehicles-move-close-toconsumer-use/.

Gaines, Linda, 2009, Lithium-ion battery recycling issues: Argonne National Laboratory, May 21, 25 p., accessed December 29, 2009, at http:www1. eere.energy.gov/ vehiclesandfuels/pdfs/merit_review_2009/propulsion_ materials/pmp_05_ gaines.pdf.

Gaines, Linda, and Cuenca, Roy, 2000, Costs of lithium-ion batteries for vehicles: Argonne National Laboratory Report ANL/ESD-42, 58 p., accessed October 8, 2009, at http://www.transpor tation.anl.gov/pdfs/ TA/149.pdf.

Gaines, Linda, and Nelson, Paul, 2009, Lithium-ion batteries—Possible materials issues: U.S. Department of Transportation, 16 p., accessed December 29, 2009, at http://www.transportation. anl.gov/pdfs/ B/583. PDF.

General Motors Company, 2010, GM builds first lithium-ion battery for Chevrolet Volt: General Motors Company, accessed July 7, 2010, at http://www.gm.com/ corporate/responsibility/environment/news/2010/ voltbattery_010710.jsp.

Green Batteries, 2009, Lithium-ion battery frequently asked questions: Green Batteries, accessed January 11, 2010, at http://www.greenbatteries. com/ libafa.html.

Green Energy News, 2008, Tibet's lithium: Green Energy News, March 23, v. 13, no. 1, 3 p., accessed October 6, 2009 at http://www.green-energy-news.com/arch/ nrgs2008/20080024.html.

Haber, Steffen, 2008, Chemetall—The lithium company: Chemetall GMBH, 28 p., accessed January 4, 2010, at http://www.rockwoodspecialties. com/rock_english/media/ ppt_files/02_04_09_Lithium_Supply_Santiago.ppt.

Hamilton, Tyler, 2009, Lithium battery recycling gets a boost: Technology Review [Massachusetts Institute of Technology], August 12, accessed September 14, 2009, at http://www.technologyreview.com/energy/23215/.

Hsiao, Eugene, and Richter, Christopher, 2008, Electric vehicles special report—Lithium Nirvana—Powering the car of tomorrow: Article in CLSA Asia-Pacific Markets, June 2, 2008, 44 p. Accessed December 2, 2009, at http://www.clsa.com/assets/files/reports/ CLAS-Jp-Electric Vehicles20080530.pdf.

American Honda Motor Co., Inc., 2010, Honda FCX Clarity—Performance: American Honda Motor Co., Inc. press release, accessed July 7, 2010, at http://automobiles.honda.com/fcx-clarity/performance.aspx.

Hybrid Cars, 2010, December 2009 dashboard—Yearend tally: Hybrid Cars, accessed January 11, 2012, at http://www.hybridcars.com/hybrid-sales-dashboard/ december-2009-dashboard.html.

Jaskula, B.W., 2008, Lithium, in Metals and minerals: U.S. Geological Survey Minerals Yearbook 2007, v. I, p. 44.1– 44.8. (Also available at http:// minerals.er.usgs.gov/minerals/pubs/commodity/lithium/myb1-2007-lithi. pdf.)

Jaskula, B.W., 2008, Lithium: U.S. Geological Survey Mineral Commodity Summaries 2008, p. 98-99. (Also available at http://minerals.er.usgs.gov/ minerals/pubs/commodity/ lithium/mcs-2008-lithi.pdf.)

Jaskula, B.W., 2009, Lithium: U.S. Geological Survey Mineral Commodity Summaries 2009, p. 94-95. (Also available at http://minerals.er.usgs.gov/ minerals/pubs/commodity/ lithium/mcs-2009-lithi.pdf.)

Jaskula, B.W., 2010, Lithium: U.S. Geological Survey Mineral Commodity Summaries 2010, p. 92-93. (Also available online at http://minerals.er.usgs. gov/minerals/pubs/ commodity/lithium/mcs-2010-lithi.pdf.)

Kanellos, 2009, Audi, Mercedes, BMW prep electric sports cars—Should Tesla, Fisker worry?: Greentech Media, September 15, accessed July 7, 2010, at http://www.greentechmedia.com/articles/read/audi-mercedes-bmw-prep-electric-sports-cars-should-tesla-fisker-worry/.

Lithium Site, 2009, Lithium characteristics: Lithium Site accessed October 6, 2009, at http://www.lithiumsite/ Home_Page.html.

Ma, Pei-hua, 2000, Comprehensive utilization of salt lake resources: Key Journal [China National Science and Technology Library], v. 15, no. 4, accessed January 2, 2009, at http://keyjournal.nstl.gov.cn/ english?qcode= dqkxjz200004002&english=1.

Mah, Paul, 2007, Three things you should already know about your lithium ion battery: Tech at Play, accessed January 13, 2010, at http://www. techatplay.com/?p=61.

Mitchell, R.L., 2006, Lithium ion batteries—High-tech's latest mountain of waste: Computerworld, accessed January 15, 2010, at http://blogs. computerworld.com/node/3285.

Murray, C.J., 2010, Chrysler to build a battery-powered electric car by 2012: Design News, March 25, accessed July 7, 2010, at http://www. designnews.com/ document.asp?doc_id=228967.

National Institute of Advanced Industrial Science and Technology, 2009, Development of a new-type lithium-air battery with large capacity—A step toward a lithium fuel cell which uses recyclable lithium: National Institute of Advanced Industrial Science and Technology, accessed August 12, 2009, at http://www.aist.go.jp/aist_e/ latest_research/ 2009/ 20090727/20090727.html.

Nelson, P.A., Santini, D.J., and Barnes, James, 2009, Factors determining the manufacturing costs of lithium-ion batteries for PHEVs: International Battery, Hybrid and Fuel Cell Vehicle Symposium, EVS24, Stavenger, Norway, May 13–16, 2009, 12 p., presentation.

New Tang Dynasty Television, 2009, Bolivia taps into lithium power: New Tang Dynasty Television, accessed January 7, 2010, at http://english. ntdtv.com/ntdtv_en/ ns_sa/2009-10-31/141415053021.html.

Nissan, 2010, Sustainable mobility comes to the United States with dedication of Nissan LEAF production site: Nissan press release, accessed July 7, 2010, at http://www.nissanusa.com/leaf-electric-car/news/ technology/ sustainnnable_mobility_comes_to_united_states#/leaf-electric-car/news/ technology/sustainable_mobility_ comes_to_united_states.

Osawa, Juro and Takahashi, Yoshio, 2010, Toshiba, Mitsubishi Motors developing electric car batteries: Wall Street Journal, July 2, accessed July 7, 2010, at http://online.wsj.com/article/ BT-CO-20100702-701638.html.

Pease, Karen, 2008, Lithium counterpoint—No shortage for electric cars: Gas2.0, accessed October 1, 2009, at http://gas2.org/2008/10/13/lithium-counterpoint-noshortage-for-electric-cars.

PlanetWatch, 2009, With cars going electric, there's this roadblock—The battery: PlanetWatch, accessed January 11, 2012, at http://www. planet watch.org/.

Simpson, A., 2006, Cost-benefit analysis of plug-in hybrid electric vehicle technology: International Battery, Hybrid and Fuel Cell Electric Vehicle Symposium and Exhibition, EVS-22, Yokohama, Japan, October 23–28,

2006, conference paper NREL/CP-540-40485, 11 p. (Also available at http://www.nrel.gov/docs/ fy07osti/40485.pdf.)

Smith, Michael, and Craze, Matthew, 2009, Lithium for 4.8 billion electric cars lets Bolivia upset market: Bloomberg, accessed January 7, 2010, at http://www.bloomberg.com/ apps/news?pid=20670001&sid=aVqbD6T3XJeM.

Tahil, William, 2007, The trouble with lithium—Implications for future PHEV production for lithium demand: Meridian International Research, 22 p., accessed October 1, 2009, at http://www.meridian-int-res.com/Projects/ Lithium/ Lithium_Problem_2.pdf.

Tahil, William, 2008, The trouble with lithium 2—Under the microscope: Meridian International Research, 54 p. accessed June 8, 2010, at http://www.meridian-int-res.com/ Projects/Lithium_Microscope.pdf.

Takeshita, Hideo, 2008, Worldwide market update on NiMH, Li-ion and polymer batteries for portable applications and HEVS: Tokyo, Japan, Institute of Information Technology, Ltd., The 25th International Battery Seminar and Exhibit, Tokyo, Japan, March 17, 2008, 26 p.

Tesla Motors, 2010, Roadster overview: Tesla Motors, accessed July 7, 2010, at http://www.teslamotors.com/ roadster/technology/battery.

Tollinsky, Norm, 2008, Xstrata boosts recycling capacity: Sudbury Mining Solutions Journal, May, accessed July 2, 2010, at http://sudburymining solutions.com/articles/ SustainableDevelopment/06-08-xstrata.asp.

Toyota, 2009, 2010 Prius plug-in hybrid makes North American debut at Los Angeles auto show: Toyota press release, accessed January 11, 2012, at http://pressroom.toyota.com/article_display.cfm?article_ id=1822.

Toxco Inc., 2009, Toxco Inc. is awarded 9.5 million from DOE to support U.S. lithium battery recycling: Toxco Inc., accessed August 18, 2010, at http://www.toxco.com/docs/ Toxco%20DOE.pdf.

Truck Trend, 2009,

U.S. Bureau of Mines and U.S. Geological Survey, 1980, Principles of a resource/reserve classification for minerals: U.S. Geological Survey Circular 831, 5 p. (Also available at http://pubs.usgs.gov/circ/ 1980/ 0831/report.pdf.)

U.S. Bureau of Mines, 1992–1995, Lithium, in Metals and minerals: U.S. Bureau pf Mines Minerals Yearbook, v. I, various pages.

U.S. Department of Energy, 2010, NYMEX light sweet crude oil futures prices: U.S. Department of Energy, Energy Information Agency, accessed daily at http://www.eia.doe.gov/emeu/international/crude2.html. [These data can be accessed at http://www.eia.doe.gov/ emeu/international/oilprice.html.]

U.S. Geological Survey, 1996–2009, Lithium, *in* Metals and minerals: U.S. Geological Survey Minerals Yearbook, v. I, various pages.

U.S. Geological Survey, 2010, Lithium statistics, *in* Kelly, T.D., and Matos, G.R., comps., Historical statistics for mineral and material commodities in the United States: U.S. Geological Survey Data Series 140, accessed January 11, 2012, at http://minerals.usgs.gov/ds/2005/140/lithium.xls.

Voelcker, John, 2010, Hyundai plans Prius-fighter hybrid hatchback, lithium battery: Green Car Reports, April 20, accessed July 7, 2010, at http://www.greencarreports.com/blog/1044360_hyundaiplans-prius-fighter-hybrid-hatchback-lithium-battery.

Wilburn, D.R., 2007, Flow of cadmium from rechargeable batteries in the United States, 1996–2007: U.S. Geological Survey Scientific Investigations Report 2007–5198, 26 p., accessed Janury 10, 2012, at http://pubs.usgs.gov/ sir/2007/5198/.

Wilburn, D.R., 2008, Material use in the United States— Selected case studies for cadmium, cobalt, lithium, and nickel in rechargeable batteries: U.S. Geological Survey Scientific Investigations Report 2008–5141, 18 p., accessed January 11, 2012, at http://pubs.usgs.gov/sir/2008/5141.

Xcel Energy Inc., 2010, Smart Grid CityTM—Building a clean energy future: Xcel Energy Inc. information sheet, 1 p., accessed November 17, 2010, at http://smartgridcity.xcelenergy.com/media/pdf/ Information-Sheet.pdf.

Xu, Jinqui, Thomas, H.R., Francis, R.W., Lum, K.R., Wang, Jingwei, and Liang, Bo, 2008, A review of proceses and technologies for the recycling of lithium-ion secondary batteries: Journal of Power Sources, v. 177, p. 512–527.

Zhang, Jiangfeng, 2009, Present situation and prospects for lithium carbonate production in China: Entrepreneur, May 26, accessed January 5, 2010, at http://www.entrepreneur. com/tradejournals/article/print/201493065.html.

In Lithium Use in Batteries:
Editors: D. R. Taylor and R. I. Young
ISBN: 978-1- 62257-037-9
© 2012 Nova Science Publishers, Inc

Chapter 2

LITHIUM-ION BATTERIES: EXAMINING MATERIAL DEMAND AND RECYCLING ISSUES[*]

Linda Gaines and Paul Nelson

ABSTRACT

Use of vehicles with electric drive, which could reduce our oil dependence, will depend on lithium-ion batteries. But is there enough lithium? Will we need to import it from a new cartel? Are there other materials with supply constraints? We project the maximum demand for lithium and other materials if electric-drive vehicles expanded their market share rapidly, estimating material demand per vehicle for four battery chemistries. Total demand for the United States is based on market shares from an Argonne scenario that reflects high demand for electric-drive vehicles, and total demand for the rest of the world is based on a similar International Energy Agency scenario. Total material demand is then compared to estimates of production and reserves, and the quantity that could be recovered by recycling, to evaluate the adequacy of supply. We also identify producing countries to examine potential dependencies on unstable regions or future cartels.

Keywords: Battery materials, lithium, recycling

---

[*] This is an edited, reformatted and augmented version of an Argonne National Laboratory, Transportation Technology R&D Center publication, dated February 2010.

INTRODUCTION

As the world energy community evaluates alternatives to petroleum for personal vehicles, every aspect of potentially important technologies must come under intense scrutiny. Technical and economic issues receive most of the attention, but material availability is important to consider whenever rapid growth is expected. Lithium-ion batteries are a very promising contributor to reducing our dependence on imported oil. But is there enough lithium? Will we need to import it from a new and unfriendly cartel? What about other battery materials? The adequacy of lithium supply was recently questioned by Tahil [1], but his conclusions were disputed by Evans [2]. In this paper, we explore the potential demand for lithium and other key battery materials if hybrids, then plug-in hybrids, and then pure electric vehicles expand their market share extremely rapidly[1]. This is not meant to be a projection, but rather an upper bound on the quantity of material that could be required. The total demand can then be compared to estimates of production and reserves, and quantities that could be recycled, to evaluate the adequacy of future supply.

Several steps are required to estimate total demand for materials. First, an estimate of total vehicle demand vs. time is combined with a scenario of percent of new sales by each technology vs. time to calculate the number of new vehicles of each type produced annually. Then, batteries are designed for each vehicle type and for each chemistry, and the percent of lithium (or cobalt, nickel, etc.) in each active material and the battery pack is calculated. The battery mass for each vehicle type is estimated, and the total annual requirement for each material calculated. Finally, materials potentially available via recycling are estimated to determine net virgin material required and compared to production and reserves.

VEHICLE DEMAND

To estimate U.S. sales of vehicles with electric drive, we extended the Energy Information Administration (EIA) projections of light vehicle sales for the United States from 2030 [3] to 2050, using a model based on Gross Domestic Product, fuel price, and projections of driving-age population by using the VISION 2007 model [4]. Only moderate growth is projected between now and 2050, and most of that is in the light truck market.

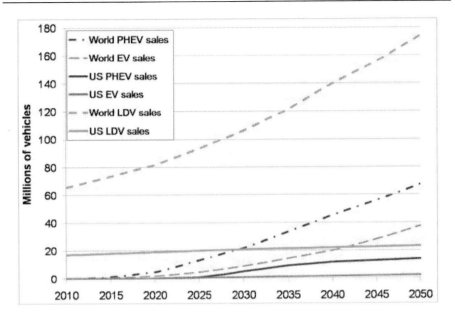

Figure 1. Light-Duty Vehicle Sales Projection to 2050.

We took the most optimistic scenario for the penetration of vehicles with electric drive into the U.S. market from the DOE Multi-Path Study (Phase 1)[5]. In this scenario, 90% of all light-duty vehicle sales are some type of electric vehicle by 2050. This is an extreme-case scenario, not a projection. It represents the maximum percent of U.S. sales that could be accounted for by hybrid vehicles like those on the road today (HEV), plug-in hybrids with different all-electric ranges (PHEVX, where X is the all-electric range in miles), and pure electric vehicles (EVs). The maximum total annual sales of vehicles with electric drive occur in 2050, with 21 million units. The cumulative total for sales of all types of electric vehicle in the United States until 2050 is 465 million vehicles.

We relied on an IEA scenario [6] for world demand. IEA is developing scenarios of what would need to be done to meet IPCC CO_2-reduction goals, based on World Bank economic and UN population projections, and the relationship between these and car ownership. The IEA scenario Characterizes an aggressive adoption of many advanced technologies; about 1.6 billion electric-drive vehicles have been built by 2050 in this scenario, with pure EVs accounting for over 20% of global sales. This is a key assumption that would cause high lithium demand. It contrasts with the Argonne optimistic scenario's

10%. Figure 1 shows our U.S. scenario, as well as IEA's projection of world LDV sales.

BATTERIES

Although the dominant chemistry used in electronics batteries today uses a mixture of nickel, cobalt, and aluminum (NCA) for the lithium salt in the active material for the cathode (positive electrode), numerous other materials are credible contenders for automotive batteries; any of them could become the major material used. We chose three promising chemistries, in addition to the current NCA graphite, to compare on the basis of material usage. These are defined in Table I. All contain lithium in a salt for the cathode active material and a lithium salt ($LiPF_6$) in the electrolyte solution. One also uses a lithium titanate salt, instead of the standard graphite, in the anode. For each battery chemistry analyzed, all materials in the electrodes and the electrolyte were tabulated to give total material required. The actual chemical formulae were used to obtain elemental percents by weight in the active compounds.

Table I. Battery Chemistries Included in the Analysis

System → Electrodes↓	NCA Graphite	LFP (phosphate) Graphite	MS (spinel) Graphite	MS TiO
Positive (cathode)	$LiNi0.8Co0.15Al0.05O_2$	$LiFePO_4$	$LiMn_2O_4$	$LiMn2O_4$
Negative (anode)	Graphite	Graphite	Graphite	$Li_4Ti_5O_{12}$

Four batteries were designed — one for each of the chemistries chosen — for each of three automobile all-electric ranges (a standard hybrid was simulated as a PHEV4). Battery designs assumed blended-mode operation, in which the engine can turn on to supply peak power demand during electric operation. Table II shows a partial breakdown of the material masses per cell. The table also shows total cell mass and numbers of cells required for each of the 12 cases.

From (1) the mass percent of each element in the active compounds and (2) the mass required of each compound in the batteries, we calculated the quantities of lithium and other materials required per battery pack. For lithium, the total is the sum of lithium from the cathode, the electrolyte, and the anode (for the cells with titanate anodes). The total requirement of lithium (on an elemental basis) for each car is shown in Table III. The electric vehicle battery requirement is based on an assumed 100-mile range. A longer range would increase both the material required and the cost to the extent that significant market penetration is unlikely. Our colleagues find that the benefit-to-cost ratio of added all-electric range for vehicles with electric drive drops very rapidly, casting doubt on the marketability of EVs with ranges greater than 100 miles [7].

Battery (and material) masses were scaled up from the designs for automobiles to ones appropriate for light trucks or sport utility vehicles, on the basis of computer runs using the Powertrain Systems Analysis Toolkit model [8], for the Multi-Path Study [9]. This is not a simple linear scale-up from the automobile masses because of the different performance features required.

Table II. Detailed Automobile Battery Composition

Parameter	Battery Type											
	NCA-G			LFP-G			LMO-TiO			LMO-G		
Vehicle Range (mi) at 300 Wh/mile	4	20	40	4	20	40	4	20	40	4	20	40
Materials Composition (g/cell)												
Cathode (+) active material	77	314	635	74	302	609	125	502	1,003	63	255	514
Anode (-) active material	51	209	423	51	208	419	83	334	669	42	170	342
Electrolyte	50	149	287	64	194	376	69	239	477	41	124	242
Total cell mass (g)	424	1,088	2,043	471	1,162	2,170	483	1,534	3,062	347	888	1,671
Cells per battery pack	60	60	60	60	60	60	60	60	60	60	60	60
Battery mass (kg)	31.2	75.9	140.1	34.6	81.6	150.2	35.6	106.2	209.1	26.1	62.6	115.4

TOTAL LITHIUM REQUIREMENTS AND RESERVES

Figure 2 shows how potential U.S. demand compares to historical world production and U.S. consumption (data from [10]). The U.S. consumption is perhaps misleading, since it only accounts for direct purchases of lithium compounds by U.S. firms and omits indirect consumption in imported batteries and products containing batteries. Once the total quantities of material required per vehicle by type were determined, they were multiplied by the annual vehicle sales by type to provide an estimate of the material demanded. Figure 2 also shows the U.S. results for lithium, assuming that all vehicles used the current NCA-Graphite chemistry. The demand is seen to rise to over 50,000 metric tons annually by 2050.

U.S. demand for lithium for automotive batteries has a very long way to go before it strains current production levels, with U.S. demand, even under this aggressive penetration scenario, not reaching current production levels until after 2030.

We then considered the potential impact of recycling on net demand for materials. Figure 2 also shows the demand curve lagged by 10 years (assumed average battery life) to approximate material that would be available for recycling if all lithium were recycled. Finally, the graph shows the difference between the gross material demand and the potentially recyclable material. This represents the net quantity of virgin material that would be required if all battery material could be recycled. Note that this curve turns over, meaning that the quantity of virgin material required actually declines after about 2035, around the point at which it reached a maximum of just under current production levels. The net demand turns around because the rate of demand growth slows. This demonstrates the importance of recycling.

Table III. Total Lithium Required per Passenger Automobile

Parameter	Battery Type															
	NCA-G				LFP-G				LMO-G				LMO-TiO			
Vehicle range (mi) at 300 Wh/mile	4	20	40	100	4	20	40	100	4	20	40	100	4	20	40	100
Vehicle type	HEV	PHEV	PHEV	EV	HEV	PHEV	PHEV	EV	HEV	PHEV	PHEV	EV	HEV	PHEV	PHEV	EV
Li in cathode (kg)	0.34	1.36	2.75	6.88	0.20	0.80	1.61	4.02	0.15	0.59	1.18	2.96	0.29	1.17	2.31	5.78
Li in electrolyte (kg)	0.04	0.10	0.20	0.51	0.05	0.14	0.26	0.53	0.03	0.09	0.17	0.43	0.05	0.17	0.34	0.84
Li in anode (kg)	0	0	0	0	0	0	0	0	0	0	0	0	0.30	1.21	2.43	6.07
Total Li in pack (kg)	0.37	1.46	2.96	7.39	0.24	0.93	1.87	4.68	0.17	0.67	1.35	3.38	0.64	2.54	5.07	12.68

There are several reasons to expect that world demand will be considerably lower than the maximum shown in Figure 3. Smaller cars with smaller batteries than IEA assumed (12–18 kWh) are likely to be used. And, it can be argued that hybrids are more attractive than battery electric cars. Further, many of the vehicles could be electric bicycles or others that require less than 10% as much lithium per vehicle. With smaller batteries and recycling, net world demand in 2050 can be kept to 4 times current production.

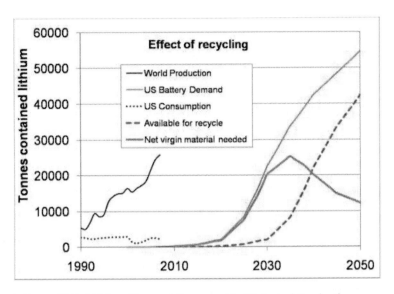

Figure 2. Future U.S. Lithium Demand Compared to Historical Production.

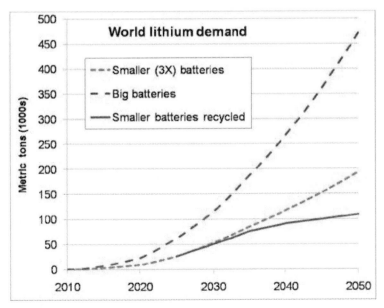

Figure 3. Future World Lithium Demand Scenarios.

We estimated cumulative battery demand for lithium under the assumption that all batteries were produced from only one chemistry. Total potential world lithium demand is shown in Table IV. (This was done for each of the four chemistries; NCA graphite is shown.) This total was then compared to several estimates of the world reserve base. The maximum demand (double the quantity shown) would occur if all batteries were made by using titanate anodes, since this chemistry uses the most lithium per battery. Only in that case does total demand exceed the USGS conservative reserve base estimate. However, by taking care with battery size and taking advantage of material that could be made available from recycling, enough lithium is available to use while we work toward an even more efficient, clean, and abundant means of supplying propulsion energy.

Table IV. World Lithium Demand and Reserves

Item	Cumulative Demand to 2050 (contained lithium, 1,000 metric tons)
Large batteries, no recycling	6,474
Smaller batteries, no recycling	2,791
Smaller batteries, recycling	1,981
USGS Reserves [11]	4,100
USGS Reserve Base [11]	11,000
Evans and others	30,000+

Figure 4 shows the locations of known lithium reserves. Chile dominates current production, with Australia second. Bolivia has huge untapped reserves, and China is rapidly developing its production capacity. The United States has limited reserves, and so it is likely to remain at least partially a materials importer, although batteries could certainly be produced here from these imported materials. The United States could supply much of its own needs and has relatively stable relationships with the major lithium-producing countries, and so significant supply problems are not anticipated.

OTHER MATERIALS

Using the same scenario and methods described earlier for lithium, we estimated the potential demand for nickel, cobalt, and aluminum for NCA-graphite batteries; iron and phosphorus for LFP batteries; manganese for either the LMOG or LMO-G; and titanium for the LMO-TiO. These quantities were then compared to USGS reserve data for each material, if appropriate. For some materials, such as iron, the quantity available is sufficiently large that another measure was used for comparison. Table V compares material availability to potential cumulative U.S. light-duty battery demand[2] to 2050 and estimates the percent that could be required. A potential constraint was identified for cobalt. If NCA-G were the only chemistry used, batteries use could impact the reserve base by 2050. Approximately 9% of the world reserve base could be required for U.S. light-duty vehicle batteries. World demand would be a factor of 4 larger. Of course, recycling — which is more likely with an expensive, scarce material — would significantly alleviate this pressure, as would the expected shift away from NCA-G to other chemistries.

The United States does not produce any cobalt, and so we must depend entirely on imports[3]. In 2006, "ten countries supplied more than 90% of U.S. imports. Russia was the leading supplier, followed by Norway, China, Canada, Finland, Zambia, Belgium, Australia, Brazil, and Morocco [12]." Cobalt is produced in many other countries as well, so it is unlikely that any one country or group could manipulate supply or price. Similarly, the United States does not produce any nickel, except for a small amount as a by-product of copper and platinum/palladium mining, so we import from Canada (41%), Russia (16%), Norway (11%), Australia (8%), and other countries (24%)[13]. Again, the diversity of producers suggests security of supply. The remaining battery materials are all abundant.

Figure 4. Sources of Lithium [11].

Table V. Comparison of U.S. Light-Duty Battery Demand to Material Availability [12,13]

Material	Availability (million tons)	Cumulative Demand	Percent Demanded	Basis
Co	13	1.1	9	World reserve base
Ni	150	6	4	World reserve base
Al	42.7	0.2	0.5	U.S. capacity
Iron/steel	1,320	4	0.3	U.S. production
P	50,000	2.3	~0	U.S. phosphate rock production
Mn	5200	6.1	0.12	World reserve base
Ti	5000	7.4	0.15	World reserve base

RECYCLING PROCESSES

Recycling can recover materials at different production stages, all the way from basic building blocks to battery-grade materials. At one extreme are smelting processes that recover basic elements or salts. These are operational now on a large scale and can take just about any input, including different battery chemistries (including Li-ion, Ni-MH, etc.) or mixed feed. Smelting takes place at high temperature, and organics, including the electrolyte and carbon anodes, are burned as fuel or reductant. The valuable metals (Co and Ni) are recovered and sent to refining so that the product is suitable for any use. The other materials, including lithium, are contained in the slag, which is now used as an additive in concrete. The lithium could be recovered by using a hydrometallurgical process [14], if justified by price or regulations.

At the other extreme, recovery of battery-grade material has been demonstrated. Such processes require as uniform feed as possible, because impurities in feed jeopardize product quality. The components are separated by a variety of physical and chemical processes, and all active materials and metals can be recovered. It may be necessary to purify or reactivate some components to make them suitable for reuse in new batteries. Only the separator is unlikely to be usable, because its form cannot be retained. This is a low-temperature process with a low energy requirement. Almost all of the energy and processing to produce battery-grade material from raw materials is saved. Large volumes are not required [15].

Battery use is expected to grow much more rapidly than any other use, so batteries will dominate lithium demand in the future. Further, growth in demand for batteries for electric-drive vehicles will dominate the battery demand [16], and so auto batteries will dominate demand for lithium after 2020. Therefore, it will be desirable to use recovered materials back in batteries.

CONCLUSION

Potential material supply constraints should be considered before embarking on an ambitious program of development for any new technology. However, shortages have often been forecast without adequate exploration or consideration of incentives rising prices might provide. For example, in 1972 the Club of Rome warned that the world would run out of gold by 1981; mercury and silver by 1985; tin by 1987; and petroleum, copper, lead, and natural gas by 1992 [17]. In the case of materials for lithium-ion batteries, it appears that even an aggressive program of vehicles with electric drive can be supported for decades with known supplies, if recycling is instituted. Of course, larger vehicles with longer ranges require more material, and so heavy reliance on pure electrics could eventually strain supplies of lithium and cobalt. Further work is required to examine recycling in more detail and to determine how much of which materials could be recovered with current or improved processes. Environmental impacts of both production and recycling processes should be quantified as well.

ACKNOWLEDGMENTS

The authors would like to thank David Howell, James Barnes, and Jerry Gibbs of the U.S. Department of Energy's Office of Vehicle Technologies for support and helpful insights. In addition, the work could not have been completed without data from Argonne staff members Margaret Singh and Steve Plotkin or without discussions with Dan Santini and Anant Vyas.

The submitted manuscript has been created by UChicago Argonne, LLC, Operator of Argonne National Laboratory ("Argonne"). Argonne, a U.S. Department of Energy Office of Science laboratory, is operated under Contract No. DE-AC02-06CH11357. The U.S. Government retains for itself, and others acting on its behalf, a paid-up nonexclusive, irrevocable worldwide license in said article to reproduce, prepare derivative works, distribute copies to the public, and perform publicly and display publicly, by or on behalf of the Government.

REFERENCES

[1] W. Tahil, "The Trouble with Lithium 2," William Tahil, Meridian International Research, Paris, France, http://www.meridian-int-res.com /Projects/Lithium_Microscope.pdf (accessed January 26, 2009), May 29, 2008.

[2] R.K. Evans, "An Abundance of Lithium," Part Two, http://www. worldlithium.com/ AN_ABUNDANCE_OF_LITHIUM_-_Part_2.html (accessed January 26, 2009), July 2008.

[3] DOE Energy Information Administration, "Assumptions to the Annual Energy Outlook 2008: Transportation Demand Module," DOE/EIA-0554(2008), http://www.eia.doe.gov/ oiaf/aeo/assumption/transportation. html (accessed January 26, 2009), released June 2008.

[4] The VISION Model, http://www.transportation.anl.gov/ modeling_ simulation /VISION/ (accessed January 29, 2009).

[5] U.S. Department of Energy, "Multi-Path Transportation Futures Study: Results from Phase 1 (March 2007)," see: http://www1.eere. energy. gov/ ba/pba/pdfs/multipath_ppt.pdf

[6] L. Fulton, IEA, personal communication with L. Gaines, Argonne National Laboratory, February 2009.

[7] D.J. Santini et al., 2009, "Where Is the Early Market for PHEVs?," World Electric Vehicle Journal, 2(4), 49–98.

[8] Powertrain Systems Analysis Toolkit model, Argonne National Laboratory, Argonne, IL, http://www.transportation.anl.gov/ modeling_ simulation/ PSAT/ (accessed January 29).

[9] S. Plotkin, and M. Singh, "Multi-Path Study Phase 2: Vehicle Characterization and Key Results of Scenario Analysis," Argonne National Laboratory, Argonne, IL, November 2008.

[10] U.S. Geological Survey, SQM, cited in Lithium, 2007 USGS Minerals Yearbook, U.S. Geological Survey, http://minerals.usgs.gov /minerals /pubs/commodity/lithium/myb1-2007- lithi.pdf (accessed January 28, 2009), August 2008.

[11] Lithium (Advance Release), Mineral Commodity Summaries, U.S. Geological Survey, http://minerals.usgs.gov/minerals/pubs/ commodity/ lithium/mcs-2008-lithi.pdf (accessed January 27, 2009), January 2008.

[12] Cobalt, 2006 Minerals Yearbook, U.S. Geological Survey, http://minerals.usgs.gov/minerals/ pubs/commodity/cobalt/myb1-2006-cobal.pdf (accessed January 28, 2009), April 2008.

[13] Nickel, U.S. Geological Survey, Mineral Commodity Summaries, http://minerals.usgs.gov/ minerals/pubs/commodity/nickel/mcs-2008-nicke.pdf (accessed January 29, 2009), January 2008.

[14] The Val'Eas Process: "Recycling of Li-ion and NiMH batteries via a Unique Industrial Closed Loop,"http://www.battery recycling.umicore. com/ download/valEasProcess Description.pdf, June 2006.

[15] S. Sloop, "Recycling Methods for Lithium-Ion and other Batteries," 13th International Battery Materials Recycling Seminar, Ft. Lauderdale, FL, March 2009.

[16] E. Anderson, "Sustainable Lithium Supplies Through 2020," (paper presented at Lithium Supply & Markets Conference, Santiago, Chile, February 2009).

[17] P. Bratby, 2008, Evidence to the House of Lords Economic Affairs Committee, www.parliament.uk/documents/upload/ EA181%20Philip% 20 Bratby.doc (accessed January 28, 2009), May 15.

End Notes

[1] We will refer to all three types as electric vehicles, or vehicles with electric drive.

[2] Assuming that all batteries were made by using only the chemistry requiring the material

[3] A fraction of current supply comes from the stockpile and recycling, but any new supply will be imported.

In Lithium Use in Batteries: ISBN: 978-1- 62257-037-9
Editors: D. R. Taylor and R. I. Young   © 2012 Nova Science Publishers, Inc

Chapter 3

LITHIUM-ION BATTERIES: POSSIBLE MATERIALS ISSUES*

Linda Gaines and Paul Nelson

ABSTRACT

The transition to plug-in hybrid vehicles and possibly pure battery electric vehicles will depend on the successful development of lithium-ion batteries. But, in addition to issues that affect performance and safety, there could be issues associated with materials. Many cathode materials are possible, with trade-offs among cost, safety, and performance. Oxides of cobalt, nickel, manganese, and aluminum in various combinations could be used, as could iron phosphates.

The anode material of choice has been graphite, but titanates may be used in the future. Similarly, different materials could be used for other parts of the cell. We consider four likely battery chemistries and estimate the quantities of all of these materials that could be required if vehicles with large batteries made significant market inroads, and we compare these quantities to world production and resources to identify possible constraints. We identify principal producing countries to identify potential dependencies on unstable regions or cartel behavior by key producers. We also estimate the quantities of the materials that could be recovered by recycling to alleviate virgin material supply restrictions and associated price increases.

---

* This is an edited, reformatted and augmented version of a Argonne National Laboratory, Transportation Technology R&D Center publication, dated June 2009.

1. INTRODUCTION

As the world energy community evaluates alternatives to petroleum for personal vehicles, every aspect of potentially important technologies must come under intense scrutiny.

Technical and economic issues receive most of the attention, but material availability is important to consider whenever rapid growth is expected — or even encouraged. Lithium-ion batteries are a very promising contributor to reducing our dependence on imported oil. But is there enough lithium? Will we need to import it from a new and unfriendly cartel? What about other battery materials?

The adequacy of lithium supply for a large battery industry was recently questioned by Tahil (2007, 2008), but his conclusions were disputed by Evans (2008). In this paper, we explore the potential demand for lithium and other key battery materials if hybrids, then plug-in hybrids, and then pure electric vehicles expand their market share extremely rapidly[1]. This is not meant to be a projection, but rather an upper bound on the quantity of material that could be required. The total demand can then be compared to estimates of production and reserves to evaluate the adequacy of future supply. Note that for this paper, demand has been estimated for U.S. vehicle use only; world demand, including that for all other battery applications, must eventually be included as well.

Several steps are required to estimate total U.S. demand for materials. These are listed below.

- Estimate total vehicle demand vs. time
- Estimate percent of new sales by each technology vs. time
- Calculate the number of new vehicles by type annually
- Design appropriate batteries for each vehicle type and for each chemistry
- Determine percent of lithium (or cobalt, nickel, etc.) in each active material and then the battery pack
- Estimate battery mass for each vehicle type
- Estimate total lithium required, by year, for each chemistry
- Estimate demand for other materials
- Estimate materials available for recycling vs. time
- Estimate net virgin material required
- Compare to production and reserves

2. VEHICLE DEMAND

For its Annual Energy Outlook 2008, the Energy Information Administration (EIA) projected light vehicle sales for the United States to 2030. Assumptions behind the EIA's transportation projections can be found on the EIA website (EIA 2008). Argonne staff extended these projections to 2050 by using a model based on Gross Domestic Product (GDP), fuel price, and projections of driving-age population. This extension was performed for the VISION 2007 model (2007).

As can be seen in Figure 1, only moderate growth is projected between now and 2050, and most of that growth is expected in the light truck market, which sees over a 50% growth in sales, while the passenger automobile market is almost stagnant.

Next, we took the most optimistic scenario for penetration of vehicles with electric drive into the U.S. market from the DOE Multi-Path Study (Phase 1) (DOE 2007). In this scenario, 90% of all light-duty vehicle sales are some type of electric vehicle by 2050 (see Figure 2). This is an extreme case scenario, not a projection. It represents the maximum percent of U.S. sales that could be accounted for by hybrid vehicles like those on the road today (HEV), plug-in hybrids with different all-electric ranges (PHEVX, where X is the all-electric range in miles), and pure electric vehicles (EVs). Plug-in hybrids are generally assumed to operate in all-electric or charge-depleting mode for the first X miles of travel, but then they run as a conventional hybrid in charge-sustaining mode when the battery state-of-charge declines to a predetermined percentage. In reality, operation in blended mode, where the engine could supply peak power during the "electric" miles, would be more efficient and allow designs with smaller and more economical batteries.

We interpolated both the total vehicle sales for passenger cars and light trucks (Figure 1) and the market shares of electric vehicle types from the Multipath Study (Figure 2) and combined them in an Excel spreadsheet to yield total numbers of vehicles sold of each type in each year, as can be seen in Figure 3. The maximum total annual sales of vehicles with electric drive occur in 2050, when they have grown to 21 million units, of which plug-in light trucks represent over 8 million units. In this scenario, sales of PHEVs are beginning to plateau, but sales of EVs are advancing, accounting for about 2.4 million new vehicles in 2050. The actual penetration of EVs will be seen as a key factor in material demand, because these vehicles require larger batteries. The cumulative total for sales of all types of electric vehicle in the United States until 2050 is 465 million vehicles.

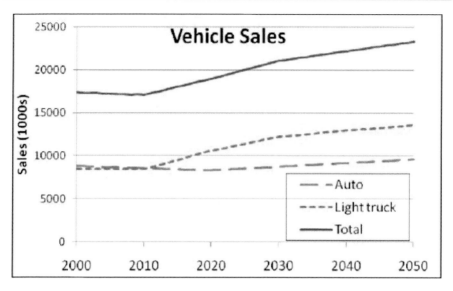

Figure 1. U.S. Light-Duty Vehicle Sales Projection to 2050.

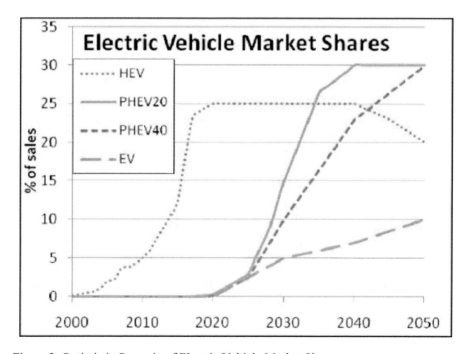

Figure 2. Optimistic Scenario of Electric Vehicle Market Shares.

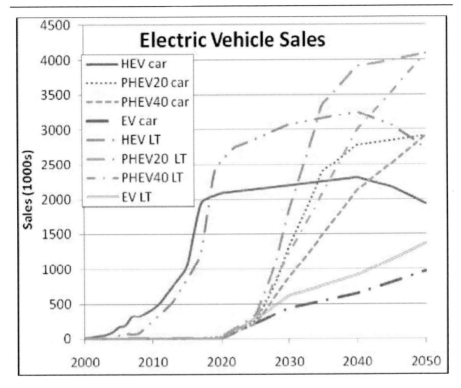

Figure 3. U.S. Electric Vehicle Sales by Type, to 2050.

Next, we needed to characterize the batteries so that we could estimate how much material would be required for each type of vehicle and then for the United States as a whole. Although the dominant chemistry used in electronics batteries today uses a mixture of nickel, cobalt, and aluminum (NCA) for the lithium salt in the active material for the cathode (positive electrode), numerous other materials are serious contenders for automotive batteries. Each has advantages and disadvantages that could eventually lead to any of these becoming the major material used. We chose three promising chemistries, in addition to the current NCA-Graphite, to compare on the basis of material usage.

These are defined in Table 1. All contain lithium in a salt for the cathode active material, and all contain a lithium salt ($LiPF_6$) in the electrolyte solution as well. One also uses a lithium titanate salt, instead of the standard graphite, in the anode.

For each battery chemistry analyzed, all materials in the electrodes and the electrolyte were tabulated to give total material required.

The actual chemical formulae were used to obtain elemental percents by weight in the active compounds, as can be seen in Table 2. For NCA-G, Li can be seen to be 6.94/96.08, or 7.2% by mass of cathode active material.

Table 1. Battery Chemistries Included in the Analysis

System Electrodes	NCA Graphite	LFP (phosphate) Graphite	MS (spinel) Graphite	MS TiO
Positive (cathode)	LiNi0.8Co0.15Al0.05 O$_2$	LiFePO4	LiMn$_2$O$_4$	LiMn$_2$O$_4$
Negative (anode)	Graphite	Graphite	Graphite	Li4Ti5O$_{12}$

Table 2. Mass of Elements in Active Compounds

Mass Element	AMU	Number per Molecule				
		NCA	LFP	MS	TiO	LiPF6
Li	6.94	1	1	1	4	1
Ni	58.69	0.8	0	0	0	0
Co	58.93	0.15	0	0	0	0
Al	26.98	0.05	0	0	0	0
O	16	2	4	4	12	0
Fe	55.85	0	1	0	0	0
P	30.97	0	1	0	0	1
Mn	54.94	0	0	2	0	0
Ti	47.88	0	0	0	5	0
F	19	0	0	0	0	6
Total Mass (AMU)		96.08	157.76	180.82	459.16	151.91

Four batteries were designed — one for each of the chemistries chosen — for each of three automobile all-electric ranges (a standard hybrid was simulated as a PHEV4). Battery designs assumed blended-mode operation. Table 3 shows a partial breakdown of the material masses per cell. The table also shows total cell mass and numbers of cells required for each of the 12 cases.

From (1) the mass percent of each element in the active compounds and (2) the mass required of each compound in the batteries, we calculated the quantities of lithium and other materials required per battery pack. For lithium, the total is the sum of lithium from the cathode, the electrolyte, and the anode (for the cells with titanate anodes). The total requirement of lithium (on an elemental basis) for each car is shown in Table 4. The electric vehicle battery requirement is based on an assumed 100-mile range. A longer range would increase both the material required and the cost to the extent that significant market penetration is unlikely. Results from other analyses, such as Tahil's, project higher demand, on the basis of assumed higher EV range. Our colleagues, who have investigated market potential of electric drive, find that the benefit-to-cost ratio of added all-electric range for vehicles with electric drive drops very rapidly, casting doubt on the marketability of EVs with ranges greater than 100 miles (Santini et al. 2009). However, if range is dropped well below 100 miles in "city electrics" to save on electric vehicle cost, then market share is estimated to drop because such vehicles meet the needs of few customers (Vyas et al. 2009). Such deductions suggest that less "EV-optimistic" scenarios are more credible.

Battery (and material) masses were scaled up from the designs for automobiles to ones that would be appropriate for light trucks or sport utility vehicles (SUVs), on the basis of computer runs using the Powertrain Systems Analysis Toolkit (PSAT) model (PSAT 2009), for the Multi-Path Study (Plotkin and Singh 2008). This is not a simple linear scale-up from the automobile masses because of the different performance features required by the different vehicle types. The battery mass for PHEV20 light trucks was estimated from Table 5 by interpolation of the PHEV10 and PHEV40 ratios of SUV battery mass to car battery mass, which are actually not very different. Similarly, 2050 ratios were obtained by extrapolation from 2045.

Table 3. Detailed Automobile Battery Composition

Parameter	Battery Type											
	NCA-G			LFP-G			LMO-TiO			LMO-G		
Vehicle Range (mi) at 300 Wh/mile	4	20	40	4	20	40	4	20	40	4	20	40
Materials Composition (g/cell)												
Cathode (+) active material	77	314	635	74	302	609	125	502	1,003	63	255	514
Anode (-) active material	51	209	423	51	208	419	83	334	669	42	170	342
Electrolyte	50	149	287	64	194	376	69	239	477	41	124	242
Total cell mass (g)	424	1,088	2,043	471	1,162	2,170	483	1,534	3,062	347	888	1,671
Cells per battery pack	60	60	60	60	60	60	60	60	60	60	60	60
Battery mass (kg)	31.2	75.9	140.1	34.6	81.6	150.2	35.6	106.2	209.1	26.1	62.6	115.4

Table 4. Total Lithium Required per Passenger Automobile

Parameter		Battery Type															
		NCA-G				LFP-G				LMO-G				LMO-TiO			
Vehicle range (mi) at 300 Wh/mile		4	20	40	100	4	20	40	100	4	20	40	100	4	20	40	100
Vehicle type		HEV	PHEV	PHEV	EV	HEV	PHEV	PHEV	EV	HEV	PHEV	PHEV	EV	HEV	PHEV	PHEV	EV
Li in cathode (kg)		0.335	1.36	2.75	6.88	0.196	0.796	1.61	4.02	0.145	0.587	1.18	2.96	0.287	1.165	2.31	5.78
Li in electrolyte (kg)		0.035	0.104	0.202	.505	0.045	0.136	0.264	.528	0.029	0.087	0.170	.425	0.049	0.167	0.335	.838
Li in anode (kg)		0	0	0	0	0	0	0	0	0	0	0	0	0.301	1.21	2.43	6.07
Total Li in battery pack (kg)		0.370	1.46	2.96	7.39	0.241	0.932	1.87	4.68	0.173	0.674	1.35	3.38	0.637	2.54	5.07	12.68

Table 5. Relative Battery Masses for Cars and Light Trucks (kg/vehicle)(Plotkin and Singh 2008)

Vehicle Category	Mass, by Type			
	HEV	PHEV10	PHEV40	EV
MID-SIZE CAR				
2015	34	46.6	92.6	316
2030	31	42.8	84.2	279
2045	32	41.3	81.7	267
MID-SIZE SUV				
2015	40	56.2	119.7	431
2030	37	52	110.8	395
2045	37	50.9	107.6	380

4. TOTAL LITHIUM REQUIREMENTS

Once the total quantities of material required per vehicle by type were determined, they could be multiplied by the annual numbers of vehicles by type to provide an estimate of the material demanded by year. Figure 4 shows the result for lithium, assuming that all vehicles used the current NCA-Graphite chemistry. The demand is seen to rise to over 50,000 metric tons annually by 2050. The demand for lithium for PHEV40 light trucks is largest by 2030, with all-electric light truck material demand second by 2040. Material demand for HEVs is almost negligible. Similar results were obtained for the other chemistries analyzed.

Next, we compared U.S. auto battery demand to world production. Future work must, of course, add demand from the rest of the world to this analysis. Figure 4 also shows how potential U.S. lithium demand compares to historical world production and U.S. consumption. The U.S. consumption is perhaps misleading, since it only accounts for direct purchases of lithium compounds by U.S. firms and omits indirect consumption in the form of imported batteries and products containing batteries. If large numbers of batteries were ever produced in the United States, the consumption curve would then reflect more realistic usage. Note that demand for lithium for automotive batteries has a very long way to go before it strains current production levels, with U.S. demand, even under this aggressive penetration scenario, not reaching current production levels until after 2030. Even if world demand were four times U.S. demand, current production levels would be sufficient to cover automotive battery demand (only) until about 2025. It is reasonable to expect the lithium production industry to be able to expand at the relatively slow rate required to meet automotive battery demand.

We then considered the potential impact of recycling on net demand for materials. Figure 4 also shows the demand curve lagged by 10 years (assumed average battery life) to approximate material that would be available for recycling if all lithium were recycled. The effect of less-optimistic assumptions and a more realistic vehicle survival function will be included in future work.

Finally, the graph shows the difference between the gross material demand and the potentially recyclable material. This represents the net quantity of virgin material that would be required if all battery material could be recycled.

Note that this curve turns over, meaning that the quantity of virgin material required actually declines after about 2035, having reached a maximum of about 20,000 metric tons per year, just under current production levels. The net demand turns around because the rate of demand growth slows.

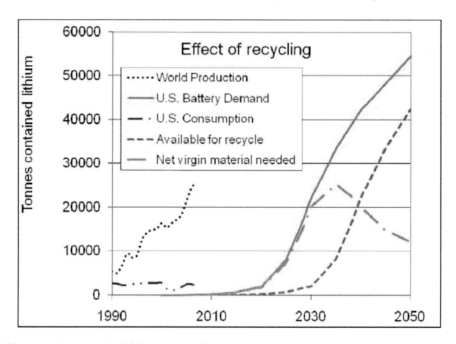

Figure 4. Future U.S. Lithium Demand Compared to Historical Production.

Of course, demand for lithium for electronics batteries — which currently makes up essentially all battery demand for lithium — must be projected forward and included. This remains to be done. Figure 5 shows that battery demand currently accounts for about 25% of world lithium production. However, batteries are by far the fastest growing use, and so future lithium demand is likely to be dominated by batteries.

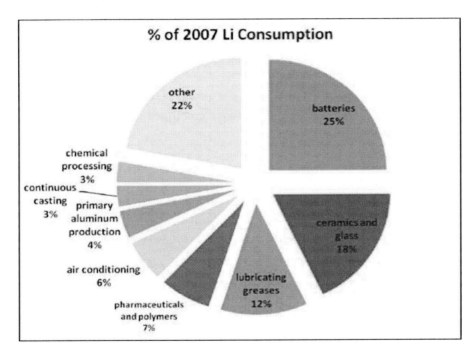

Figure 5. Current World Lithium Markets (USGS 2008a).

We also estimated cumulative battery demand for lithium and other materials for light-vehicle batteries, under the assumption that 100% of all batteries were produced from only one chemistry. Total (gross) potential lithium demand for the four chemistries is shown in Figure 6. (This was done for each of the four chemistries in turn, so the total demand numbers should not be added.) This total was then compared to United States Geological Survey (USGS) estimates of the world reserve base, which are considerably lower than recent estimates by experts (Evans 2008). USGS estimates are shown in Table 6.

The maximum demand would occur if all batteries were made by using titanate anodes, since this chemistry uses the most lithium per battery. But even in that case, total demand is about 1.8 million metric tons, compared to world reserves and reserve base of 4 and 11 million metric tons, respectively. (The USGS definitions of reserve and reserve base are provided in the Appendix.) Even when our U.S. estimates are multiplied by a factor of 4 to account for world demand, it appears that there is enough lithium available to use while we work toward an even more efficient, clean, and abundant means of supplying propulsion energy.

Table 6 also lists the locations of current lithium production and known reserves. Chile dominates current production, with Australia second. Bolivia has huge untapped reserves, and China is rapidly developing its production capacity.

The United States has very limited reserves, and so it is likely to always be a materials importer, although batteries could certainly be produced here from these imported materials.

The United States has relatively stable relationships with the major lithium-producing countries, and so significant supply problems are not anticipated at present.

Table 6. Lithium Production and Reserve Statistics (adapted from USGS 2008b)

World Mine Production, Reserves,[a] and Reserve Base[a]:				
	Mine production		Reserves	Reserve base
	2006	2007[e]		
United States	W	W	38,000	410,000
Argentina[e]	2,900	3,000	NA	NA
Australia[e]	5,500	5,500	160,000	260,000
Bolivia	—	—		5,400,000
Brazil	242	240	190,000	910,000
Canada	707	710	180,000	360,000
Chile	8,200	9,400	3,000,000	3,000,000
China	2,820	3,000	540,000	1,100,000
Portugal	320	320	NA	NA
Russia	2,200	2,200	NA	NA
Zimbabwe	600	600	23,000	27,000
World total (rounded)	23,500	25,000	4,100,000	11,000,000

[a] See Appendix for definitions.
[e] Estimated.

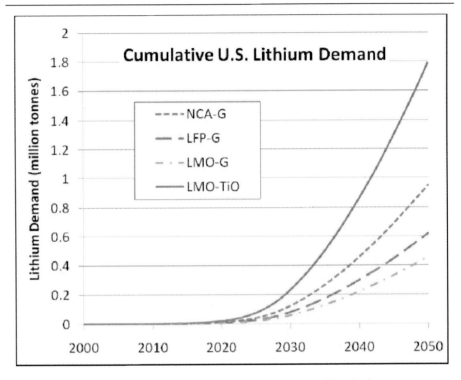

Figure 6. Cumulative U.S. Lithium Demand for Four Battery Chemistries.

5. OTHER MATERIALS

We also estimated cumulative demand for other materials that could be required for electrodes of lithium-ion batteries. Using the same scenario and methods described earlier for lithium, we prepared the potential demand for nickel, cobalt, and aluminum for NCA-graphite batteries; iron and phosphorus for LFP batteries; manganese for either the LMO-G or LMO-G; and titanium for the LMO-TiO. Figure 7 shows the cumulative demand for these materials, again assuming that all U.S. light-duty electric vehicle batteries were made by using only the chemistry requiring the material.

These quantities were then compared to USGS reserve data for each material, if appropriate. For some materials, such as iron, the quantity available is sufficiently large that another measure was used for comparison. Table 7 compares material availability to potential cumulative U.S. light-duty battery demand to 2050 and estimates the percent that could be required. A potential constraint was identified for one material. If NCA-G were the only chemistry used, cobalt use could make a dent in the reserve base by 2050. Approximately 9% of the world reserve base could be required by 2050 for U.S. light-duty vehicle batteries. Of course, recycling — which is more likely with an expensive, scarce material — would significantly alleviate this pressure.

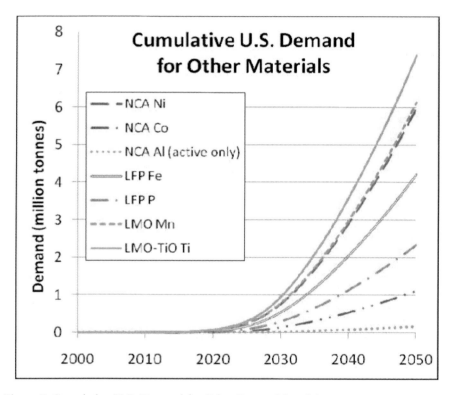

Figure 7. Cumulative U.S. Demand for Other Battery Materials.

The United States does not produce any cobalt, and so we must depend entirely on imports[2]. In 2006, "ten countries supplied more than 90% of U.S. imports. Russia was the leading supplier, followed by Norway, China, Canada, Finland, Zambia, Belgium, Australia, Brazil, and Morocco (USGS 2008c)." Cobalt is produced in many other countries as well, so it is unlikely that any one country or group could manipulate supply or price. Similarly, the United States does not produce any nickel, except for a small amount as a by-product of copper and platinum/palladium mining, so we import from the followingproducers: Canada, 41%; Russia, 16%; Norway, 11%; Australia, 8%;

economically available within planning horizons beyond those that assume proven technology and current economics. The reserve base includes those resources that are currently economic (reserves), marginally economic (marginal reserves), and some of those that are currently sub-economic (sub-economic resources). The term "geologic reserve" has been applied by others generally to the reserve-base category, but it also may include the inferred-reserve-base category; it is not a part of this classification system.

ACKNOWLEDGMENTS

The authors would like to thank David Howell, James Barnes, and Jerry Gibbs of the U.S. Department of Energy's Office of Vehicle Technologies for support and helpful insights. In addition, the work could not have been completed without data from Argonne staff members Margaret Singh and Steve Plotkin or without discussions with Dan Santini and Anant Vyas.

The submitted manuscript has been created by UChicago Argonne, LLC, Operator of Argonne National Laboratory ("Argonne"). Argonne, a U.S. Department of Energy Office of Science laboratory, is operated under Contract No. DE-AC02-06CH11357. The U.S. Government retains for itself, and others acting on its behalf, a paid-up nonexclusive, irrevocable worldwide license in said article to reproduce, prepare derivative works, distribute copies to the public, and perform publicly and display publicly, by or on behalf of the Government.

REFERENCES

Bratby, P., 2008, Evidence to the House of Lords Economic Affairs Committee inquiry into "The Economics of Renewable Energy," www. parliament.uk/documents/upload/EA181%20Philip%20Bratby.doc (accessed January 28, 2009), May 15.

DOE: U.S. Department of Energy

DOE 2007, "Multi-Path Transportation Futures Study: Results from Phase 1 (March 2007)," see: http://www1.eere.energy.gov/ba/pba/ pdfs/multipath_ppt.pdf.

EIA: DOE Energy Information Administration.

EIA, 2008, "Assumptions to the Annual Energy Outlook 2008: Transportation Demand Module," DOE/EIA-0554(2008), http://www.eia.doe.gov/oiaf/aeo/assumption/transportation.html (accessed January 26, 2009), released June.

EIA, 2008, "Assumptions to the Annual Energy Outlook 2008: Transportation Demand Module," DOE/EIA-0554(2008), http://www.eia.doe.gov/oiaf/aeo/assumption/transportation.html (accessed January 26, 2009), released June.

Evans, R.K., 2008, "An Abundance of Lithium," Part Two, http://www.worldlithium.com/AN_ABUNDANCE_OF_LITHIUM_-_Part_2.html (accessed January 26, 2009), July.

Kromer, M.A., and J.B. Heywood, 2007, "Electric Powertrains: Opportunities and Challenges in the U.S. Light-Duty Vehicle Fleet," Laboratory for Energy and the Environment, publication No. LFEE 2007-03 RP, Massachusetts Institute of Technology, Cambridge, MA, May.

Plotkin, S., and M. Singh, 2008, "Multi-Path Study Phase 2: Vehicle Characterization and Key Results of Scenario Analysis," (to be published), Argonne National Laboratory, Argonne, IL, November.

PSAT: Powertrain Systems Analysis Toolkit model

PSAT 2009, Argonne National Laboratory, Argonne, IL, http://www.transportation.anl.gov/modeling_simulation/PSAT/ (accessed January 29).

Santini, D.J., et al., 2009, "Where Is the Early Market for PHEVs?," World Electric Vehicle Journal, Vol. 2, No. 4, pp. 49–98.

Tahil, W., 2007, "The Trouble with Lithium," http://www.evworld.com / library/lithium_shortage.pdf (accessed January 26, 2009), January.

Tahil, W., 2008, "The Trouble with Lithium 2," William Tahil, Meridian International Research, Paris, France, http://www.meridian-int-res.com/Projects/Lithium_Microscope.pdf (accessed January 26, 2009), May 29.

USGS: U.S. Geological Survey.

USGS 2008a, SQM, cited in Lithium, 2007 USGS Minerals Yearbook, U.S. Geological Survey, http://minerals.usgs.gov/minerals/pubs/ commodity/lithium/myb1-2007-lithi.pdf (accessed January 28, 2009), August.

USGS 2008b, Lithium (Advance Release), Mineral Commodity Summaries, U.S. Geological Survey, http://minerals.usgs.gov/minerals/pubs/commodity/lithium/mcs-2008-lithi.pdf (accessed January 27, 2009), January.

USGS 2008c, Cobalt, 2006 Minerals Yearbook, U.S. Geological Survey, http://minerals.usgs.gov/minerals/pubs/commodity/cobalt/myb1-2006-cobal.pdf (accessed January 28, 2009), April.

USGS 2008d, Nickel, U.S. Geological Survey, Mineral Commodity Summaries, http://minerals.usgs.gov/minerals/pubs/commodity nickel/mcs-2008-nicke.pdf (accessed January 29, 2009), January.

VISION 2007, The VISION Model, http://www.transportation.anl.gov/modeling_simulation/VISION/ (accessed January 29, 2009).

Vyas, A.D., D.J. Santini, and L.R. Johnson, 2009, "Plug-In Hybrid Electric Vehicles' Potential for Petroleum Use Reduction: Issues Involved in Developing Reliable Estimates," Transportation Research Board 88th Annual Meeting, Paper No. 09-3009, Washington, D.C., January 11–15, 2009.

End Notes

[1] We will refer to all three types as electric vehicles, or vehicles with electric drive.

[2] A fraction of current supply comes from the stockpile and recycling, but any new supply will be imported.

In Lithium Use in Batteries:
Editors: D. R. Taylor and R. I. Young

ISBN: 978-1- 62257-037-9
© 2012 Nova Science Publishers, Inc

Chapter 4

REDUCING FOREIGN LITHIUM DEPENDENCE THROUGH CO-PRODUCTION OF LITHIUM FROM GEOTHERMAL BRINE*

Kerry Klein and Linda Gaines

ABSTRACT

Following a 2009 investment of $32.9 billion in renewable energy and energy efficiency research through the American Recovery and Reinvestment Act, President Obama in his January 2011 State of the Union address promised deployment of one million electric vehicles by 2015 and 80% clean energy by 2035.

The United States seems poised to usher in its bright energy future, but in reality, industry supply chains still rely on foreign sources for many key feedstock materials. In particular, 43% of the lithium consumed domestically is imported, primarily from Chile, Argentina and China, and in 2010, only one company produced lithium from U.S. resources (USGS, 2011).

Geothermal brines of the Imperial Valley resources of Southern California have been shown to be especially enriched in lithium but today remain an untapped resource.

Producing lithium battery feedstocks at geothermal production facilities could not only provide the U.S. with much-needed lithium

* This is an edited, reformatted and augmented version of an Argonne National Laboratory, Transportation Technology R&D Center publication, dated October 2011.

products and by-products, but could provide millions of dollars in added revenue to geothermal developers.

By providing lithium reserve estimates, Imperial Valley production potential and forecasts of the future of the electric vehicles industry, this study aims to relate the imperative of domestic lithium production to the vast potential of U.S. geothermal resources and showcase the benefits of industry adoption of lithium co-production at viable geothermal power plants.

Keywords: Mineral extraction, zinc, silica, strategic metals, Imperial Valley, lithium ion batteries, electric-drive vehicles, battery recycling

1. INTRODUCTION

The vast potential for mineral recovery from geothermal fluids has been recognized for decades.

Beginning in the 1970s with potash extraction from Cerro Prieto, Mexico, metal and mineral extraction techniques have been demonstrated at a wide array of resources and today represent a promising means of producing metals and materials imperative to the rapidly developing clean energy industries in the United States.

Today, 43% of the lithium consumed domestically is imported, primarily from Chile, Argentina and China, and in 2010, only one company produced lithium from U.S. resources (USGS, 2011).

From 1996 to 2005, apparent lithium consumption in the U.S. more than tripled (Wilburn, 2008) and is projected to increase dramatically with increased manufacture and demand for lithium ion batteries for use in electric-drive vehicles. Actual imports and consumption are higher because we also import lithium-containing batteries and devices containing batteries for use.

With 4,000,000 tons of estimated resources, 38,000 of which are currently economically attractive (see Table 1), there is no reason for the U.S. to be as reliant on foreign lithium sources as it is today. Decades of research and an ongoing lithium extraction pilot in Southern California show promise for this industry, and if proven to be technically feasible and financially competitive on a commercial scale, the adoption of co-production of lithium at geothermal plants has the potential to significantly reduce our dependence on foreign lithium sources.

Table 1. Global Lithium Production, Reserves and Resources (USGS, 2011)

Country	2010 Production (metric tons)	Reserves (metric tons)	Resources (metric tons)
United States	N/A	38,000	4,000,000
Argentina	2,900	850,000	2,600,000
Australia	8,500	580,000	630,000
Brazil	180	64,000	1,000,000
Canada	0	N/A	360,000
Chile	8,800	7,500,000	7,500,000
China	4,500	3,500,000	5,400,000
Portugal	0	10,000	N/A
Zimbabwe	470	23,000	N/A
Bolivia	N/A	N/A	9,000,000
Congo	N/A	N/A	1,000,000
Serbia	N/A	N/A	1,000,000
World Total	25,350	12,565,000	32,490,000

2. OVERVIEW OF METAL EXTRACTION

Geothermal fluid, or "geofluid," is the hot brine pumped from depth to the surface at geothermal production facilities. In a typical production scenario, after energy is extracted, the now-cooled "spent brine" is reinjected into the reservoir in a continuous loop. Mineral extraction methods are applied to spent brine, downstream from power production and just prior to reinjection into the reservoir. The symbiosis of material extraction facilities at operating geothermal power plants is ideal because the infrastructure for fluid flow-through and processing is largely already in place, and the royalties from the sale of by-products can provide an added revenue stream for the developers.

While the lithium dominating the market today was produced by brine evaporation and hard-rock mining methods, recovery from geothermal fluids involves a variety of chemical extraction techniques which operate quickly and without losing substantial fluid volume. Lithium can be extracted from geofluids using ion exchange resins or through the use of sorbents and stripping solutions (Bourcier et al., 2005).

Some of the earliest attempts at mineral extraction targeted silica, SiO_2, with the goal of producing high-purity colloidal silica and other commercial silica products. Other materials that have been attempted include zinc and potash. A variety of pilot extraction plants have come online in the past few decades in Nevada and California, many of which were supported by partnerships with the Department of Energy, Lawrence Livermore and Brookhaven National Laboratories, but none have enjoyed commercial success.

The closest that mineral recovery has come to commercial sustainability in the U.S. was a zinc extraction operation that came online at a CalEnergy facility in the Imperial Valley, CA. As a demonstration plant in the late 90s, this operation produced over 41,000 lbs of high-grade metallic zinc over the course of 10 months (Kagel, 2008). In the early 2000s, this demonstration scaled up to a commercial plant expected to annually produce 30,000 metric tons of 99.99% pure zinc (Clutter, 2000), but the project experienced operational and financial difficulties and was shut down within two years in 2004 (Kagel, 2008).

Today, another attempt at metal extraction is underway in the Imperial Valley. Start-up company Simbol Materials has a site access agreement with CalEnergy to operate a pilot lithium extraction plant at Elmore, where zinc extraction was attempted in 2002. If the pilot is successful, Simbol hopes to begin extracting lithium at a commercial scale in the Imperial Valley, validating these methods as a potentially significant supply of strategic metals.

3. GEOCHEMISTRY OF GEOTHERMAL BRINES

Due to variations in depth, fluid source, host rock and rock-fluid interactions in different geological regimes, geofluids exhibit a wide distribution of properties. Acidities range from pH 5 to 9 (Bourcier et al, 2005), while economically viable temperatures range from ~200°F (~93°C) and higher (Kagel, 2008). Total dissolved solids (TDS) typically range from 1,000 to 300,000 parts per million (ppm) (Bourcier et al, 2005), but have been shown to vary according to well flow rate and from location to location within a resource (Maimoni, 1982). Brine chemistries are also known to differ in pre-flash and post-flash fluids from the same well (Maimoni, 1982).

The most common elements in geothermal brines are consistently sodium, calcium, potassium and chlorine, in concentrations ranging from tens to tens of thousands of ppm. Present in significant but lesser amounts are other alkali metals, alkaline earth metals, and halides. Lithium concentrations can vary by orders of magnitude and will be discussed in more detail in the next section.

Ore and base metals such as iron, manganese, zinc, lead and copper are variably present, while precious metals such as silver, gold, platinum and palladium are typically present only in concentrations of parts per billion and commonly escape detection entirely. The highest concentrations of base and precious metals tend to be in hyper-saline brines. Silica is ubiquitous in geofluids, typically in the range of 400-800 ppm (Gallup, 1998). Table 2 shows representative elemental concentrations from a selection of geothermal resources.

4. THE POTENTIAL FOR LITHIUM RECOVERY IN THE IMPERIAL VALLEY [1]

Some of the most mineral-rich and highly studied geothermal reservoirs in the United States are located in the Imperial Valley of Southern California. With TDS ranging from 200,000-250,000 ppm in the deeper brine reservoir (Schultze & Bauer, 1982), Salton Sea brines in particular contain a veritable treasure trove of valuable metals and minerals.

The prospect of estimating a total lithium resource in the Imperial Valley is contentious. It is easier, instead, to evaluate the potential for lithium production at a representative Salton Sea power plant producing 40MW of electricity. For the purposes of this paper, a conservative estimate of lithium reserves will be proposed, with the caveat that there is tremendous potential for higher returns.

As can be seen in Table 2, lithium in the Imperial Valley resource areas is present in much greater amounts to orders of magnitudes above those in other resources. In the Salton Sea in particular, Schultze & Bauer (1982) in published a "typical" lithium concentration of 170ppm, and Maimoni (1982) attained a broad range of 117-245ppm from two wells. In this calculation, the outliers presented here will be disregarded and average lithium concentration will be considered to be 170ppm.

Due to the commonly proprietary nature of flow tests, some discrepancies exist regarding the flow rate needed to maintain 40MW of electrical production. In a 2010 press release, Ram Power announced that its Orita #2 well in the Imperial Valley had sustained a steam flow rate of 500,000 lbs/hr and was estimated to provide 8-10 MW of power. Scaling this up linearly, one could expect a flow rate of approximately 2 million lbs/hr of fluid at a typical 40MW geothermal plant.

Table 2. Global Lithium Production, Reserves and Resources (USGS, 2011)

Species (ppm)	Salton Sea, CA	Brawley, CA	Imperial Valley, CA	Coso, CA	Dixie Valley, NV	Roosevelt, UT	Mississippi Salt Dome Basin, USA	Wairakei, New Zealand	Asal, Djibouti	Milos, Greece	Cheleken Peninsula, Turkmenistan
Li	194	219	327	45	2	27	N/A	13	N/A	81	215
Na	53,000	47,600	65,500	2,850	407	2,190	59,200	1,250	29,000	31,500	76,140
K	16,700	12,600	12,450	927	64	400	538	210	5,500	9,500	409
Rb	170	67	N/A	N/A	N/A	N/A	N/A	3	N/A	N/A	N/A
Cs	20	19	N/A	N/A	N/A	N/A	N/A	3	N/A	N/A	14
Mg	33	114	400	<0.35	0	0	1,730	N/A	30	4	54
Ca	27,400	21,500	23,700	75	8	10	36,400	12	18,500	4,380	19,710
Sr	411	1,043	N/A	3	0	1	1,100	N/A	N/A	70	400
Ba	203	992	2,260	N/A	N/A	N/A	61	N/A	N/A	37	235
Fe	1,560	3,733	4,160	N/A	<0.01	N/A	298	<0.01	N/A	19	2,290
Al	2	1	4	N/A	2	N/A	N/A	N/A	N/A	N/A	N/A
Cl	151,000	134,000	131,000	5,730	438	3,650	158,200	2,210	81,000	65,400	157,000

A 2009 report by the California Division of Oil, Gas and Geothermal Resources reports average flow rates of 3-4 million lbs/hr, for a typical 40MW plant. As a highly conservative estimate, the former figure of 2 million lbs/hr will be used.

The process for extracting lithium from the brine has yet to be validated on a large-scale, long-term basis. While an ideal case would yield >90% lithium recovery, in this scenario it will be estimated at 75%. The conversion factor from lithium metal to lithium carbonate is stoichiometrically rounded to 5.32.

With all of these assumptions in place, a typical 40MW plant in the Imperial Valley resource areas could expect to produce at least 5,400 metric tons of lithium carbonate annually from extracted lithium chloride. If the value of lithium carbonate remains at its current market price of ~$5,000/metric ton (Ehren, 2009) the minimum gross annual revenue from a single 40MW plant would start at ~$28 million. Because published data indicate the likelihood of higher flow rates and higher lithium concentrations in the brine, actual produced lithium and revenues could prove to be much higher than expected. Wild cards in this scenario are the lithium recovery rate and purity of produced lithium carbonate, which have yet to be validated.

5. THE CURRENT STATE OF LITHIUM PRODUCTION

Lithium and lithium compounds are utilized in a wide variety of industrial products. Industry-ready lithium materials include lithium compounds and lithium ore concentrates. In 2009, global end-use markets for lithium were as follows: 31% for ceramics and glass, 23% for batteries, 9% for lubricating greases, 6% for air treatment, 6% for primary aluminum production, 4% for continuous casting, 4% for rubber and thermoplastics, 2% for pharmaceuticals, and 15% for other uses. A small amount of lithium carbonate is recovered domestically through the recycling of lithium batteries (USGS, 2011).

The utilization of lithium in the battery industry has enormous growth potential. Both rechargeable and non-rechargeable lithium batteries are used in a variety of portable electronics and tools, while rechargeable lithium ion batteries are being developed for hybrid and electric vehicles. Most lithium minerals are used directly as ore concentrates, but lithium carbonate and lithium hydroxide are the prevalent battery feedstock materials.

Chile dominates the world lithium market. Other major producers of lithium compounds and ore concentrates include Argentina, China, Australia,

Canada, Portugal, Zimbabwe, and the United States (USGS, 2011). Table 2 provides salient statistics from major global lithium producers.[2]

Domestically, only one company is currently producing lithium from U.S. soil. Chemetall Foote Corp., its sole operation a brine evaporation unit at Silver Peak, NV, does not disclose the quantities of lithium it produces. In 2010, the apparent domestic lithium consumption was 1,000 metric tons (USGS, 2011) but does not include consumption of imported batteries. The U.S. has nevertheless become the world leader in exporting downstream lithium compounds, which it produces domestically from imported carbonate, lithium hydroxide and lithium chloride. In the last decade, the U.S. has imported >50% of the lithium it consumed, but as of 2010 the net import reliance decreased to ~43% as a result of reduction in consumption due to the recent economic downturn. Not taken into account in these figures are direct battery imports, which are undoubtedly significant but not easily documented. Several companies have staked claims in Nevada to explore the potential of lithium-bearing aquifers but none have reached production phase (USGS, 2009).

Conventional means of producing lithium include brine evaporation and hard-rock extraction from the mineral spodumene. Brine evaporation, by far the most common method, employs a process known as Lime Soda Evaporation in which subsurface brine is pumped into evaporation ponds. The brine is treated with lime and soda ash and over a period of 12 to 18 months evaporates so that lithium chloride and a variety of other salts crystallize (Tahil, 2007).

Spodumene, a lithium-bearing silicate with chemical formula $LiAlSi_2O_6$, is typically harvested from large pegmatite veins. In its purest form, spodume forms highly sought after pink and green gemstones (Mindat, 2011). Until 1997, most lithium carbonate produced was obtained from spodumene mining, but the entry of far-cheaper brine evaporation drove most spodumene mining operations out of the market (Tahil, 2007).

Despite their proven technologial feasibilities, these two methods introduce a variety of environmental concerns that could be eliminated through lithium extraction from geofluids. Large-scale evaporation requires immense swaths of land and long-term solar exposure, produces enormous quantities of waste salts, and introduces the question of water rights in areas of freshwater scarcity. Hard-rock spodumene mining is invasive and produces large quantities of mine waste. In the case of geofluid co-production, the only land use and disruption occur during the development of the power plant and drilling of injection/production wells, and water loss is minimal as produced fluid is reinjected into the reservoir after energy extraction and processing take place.

6. LITHIUM ION BATTERIES IN ELECTRIC VEHICLES: A GROWING MARKET

Lithium-ion batteries for use in electric vehicles (EVs) are a very promising contributor to reducing our dependence on imported oil. But is there enough lithium? Will we need to import it from a new and unfriendly cartel? The market for lithium is likely to be dominated by its use in batteries, as can be seen in Figure 1 (Anderson, 2011), and automotive batteries are by far the fastest-growing type. We explore the potential demand for lithium in vehicle batteries if hybrid vehicles like those on the road today, plug-in hybrids with different all-electric ranges, and pure electric vehicles expand their market share extremely rapidly. This is not a projection, but rather an upper bound on the quantity of material that could be required. The total demand is then compared to estimates of production and reserves, and quantities that could be recycled, to evaluate the adequacy of future supply.

First, an estimate of total vehicle demand vs. time was combined with a scenario of percent of new sales by each technology vs. time to calculate the number of new vehicles of each type produced annually. To estimate U.S. sales of vehicles with some form of electric drive, we extended the Energy Information Administration (EIA) projections of light vehicle sales for the United States from 2030 (DOE/EIA, 2008) to 2050, using a model based on Gross Domestic Product, fuel price, and projections of driving-age population by using the VISION 2007 model (Argonne, 2008). We took the most optimistic scenario for penetration into the U.S. market from the DOE Multi-Path Study (Phase 1) (DOE, 2007). In this scenario, 90% of all light-duty vehicle sales are some type of electric vehicle by 2050. This represents the maximum percent of U.S. sales that could be accounted for by. Upper-bound world demand was also estimated. We relied on an IEA scenario (IEA, 2009) for world demand. IEA is developing scenarios of what would need to be done to meet IPCC CO_2-reduction goals. This scenario characterizes an aggressive adoption of many advanced technologies; about 1.6 billion electric-drive vehicles have been built by 2050 in this scenario, with pure EVs accounting for over 20% of global sales. This is a key assumption that would cause high lithium demand.

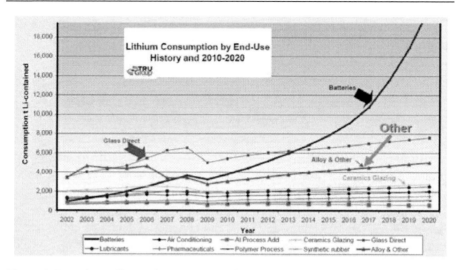

Figure 1. Batteries Will Dominate Future Lithium Demand (borrowed with permission from TRU, 2011).

Batteries were designed for each vehicle type, the battery masses estimated, and the total annual lithium requirement calculated, assuming that all vehicles used the current NCA-Graphite chemistry. Finally, materials potentially available via recycling were estimated to determine net virgin material required, and compared to production and reserves. Table 3 (Gaines et al., 2010) shows how total cumulative world demand for lithium up to 2050 would compare to reserve estimates for the IEA scenario and a similar scenario using batteries a factor of 3 smaller, both recycled and not. The economic reserves would not be used up, and recycling could extend them significantly.

Table 3. World Lithium Demand and Reserves (Gaines et al., 2010)

	Cumulative Demand to 2050 (Contained Lithium, 1000 Metric Tons)
Large batteries, no recycling	6,474
Smaller batteries, no recycling	2,791
Smaller batteries, recycling	1,981
	Reserve Estimates
USGS Reserves	9,900
USGS World Resource	25,500
Other Reserve Estimates	30,000+

One reason for increasing domestic U.S. production of lithium would be to become less dependent on imported raw materials. Under our optimistic EV scenario, the U.S. would require a total of about one million tons of lithium with no recycling, and about half that if all the material could be recycled after 10 years of use. The currently economic reserves in the U.S. are 38,000 tons (USGS, 2011), clearly insufficient, but the total resource (available at higher cost) is about 4 million tons, which could eliminate our lithium imports. However, the price would need to be higher. How would that impact the cost of the vehicles that would be using the material in their batteries? The Chevrolet Volt and the Nissan Leaf, the 2 electric-drive vehicles now available in the U.S., contain about 2 kg and 4 kg of lithium in their battery packs, respectively. At the current market price of ~$5000/ton of lithium carbonate (Ehren, 2009), that raw material costs $50-$100 per vehicle. When this is compared to the vehicle prices of $30,000-$40,000, it is clear that even a doubling of the cost of lithium would not be a significant perturbation on the vehicle price. Therefore, a cost increase (if it were to happen) caused by using domestic suppliers could be tolerated by the vehicle market.

If 5,400 metric tons of lithium carbonate were produced annually from the Salton Sea, 250,000- 500,000 cars using 2-4 kg of lithium could use this domestic resource for their batteries. At that rate, only 2-4 years' production would meet President Obama's one million vehicle goal.

7. DISCUSSION

The advantages of co-produced lithium over conventional lithium production are manifold. If the electric vehicles industry is to expand as it has the potential to, and as President Obama has promised, the foreign dependence of our lithium supply chain cannot be overlooked. The U.S. possesses an abundance of lithium, but even the little that the U.S. consumes has the potential to be produced domestically, providing political, environmental and financial benefits to U.S. consumers and geothermal power developers.

The economic and scientific potential of co-production from geothermal brines has been abundantly demonstrated in scientific literature and in decades of on-site demonstrations. With the potential to provide at least 5,400 metric tons of lithium carbonate and over $28 million in gross revenues to a single 40MW plant in the many resource areas of the Imperial Valley, CA, the promise of this technology is clear. Combining this potential with the environmental responsibility of the technology and the promise of greater political stability as compared to conventional production methods, this technology will be a key player in the energy future of the United States.

ACKNOWLEDGMENTS

This work was funded by the Department of Energy's Geothermal Technologies and Vehicles Technologies Programs. Thanks are extended from primary author Kerry Klein to Jay Nathwani, John Ziagos, Carol Bruton and Stephen Harrison for their encouragement and creative and technical input.

REFERENCES

Argonne, 2008: Argonne National Laboratory, 2008. "The VISION Model." http://www.transportation.anl.gov/modeling.simulation/VISION/. Accessed January 2009.

Gaines, L., P. Nelson, 2010. "Lithium-Ion Batteries: Examining Material Demand and Recycling Issues." The Minerals, Metals and Materials Society 2010 Annual Meeting and Exhibition, Seattle, WA, USA.

Bourcier, W.L., M. Lin, G. Nix, 2005. "Recovery of minerals and metals from geothermal fluids." Lawrence Livermore National Laboratory, Livermore, CA, USA, 18pp.

U.S. Department of Energy/Energy Information Administration (DOE/EIA), 2008. "Assumptions to the Annual Energy Outlook 2008: Transportation Demand Module." DOE/EIA-0554(2008), http://www.eia.doe.gov/oiaf/aeo/assumption/transportation.html.

U.S. Department of Energy (DOE), 2007. "Multi-Path Transportation Futures Study: Results from Phase 1." http://www1.eere.energy. gov/ba/pba/ pdfs/ multipath ppt.pdf.

Ehren, P., 2009. "The Lithium Site." http://www.lithium Market.html. Accessed May 2011.

Gallup, D.L., 1998. "Geochemistry of geothermal fluids and well scales, and potential for mineral recovery." Ore Geology Reviews 12, pp. 225-236.

IEA, 2009: L. Fulton, IEA, personal communication with L. Gaines, Argonne National Laboratory, February 2009.

Johnson, E.A., 2009. "The Preliminary 2008 Annual Report of California Oil, Gas and Geothermal Production: Summary of Geothermal Operations."

California Division of Oil, Gas, and Geothermal Resources (DOGGR), Sacramento, CA, USA. ftp://ftp.consrv.ca.gov/pub/oil annual reports/2008/0109geofin 08.pdf.

Kagel, Alyssa, 2008. "The State of Geothermal Energy, part II: Surface Technology." Geothermal Energy Association, Washington, DC, USA, 78pp. http://www.geo-energy.org/ reports/Geothermal%20Technology% 20-%20Part%20II%20(Surface).pdf.

Maimoni, A., 1982. "Minerals recovery from Salton Sea geothermal brines: A literature review and proposed cementation process." Geothermics 11, pp. 239-258.

Mindat, 2011. "Spodumene." http://www.mindat.org/min-3733.html. Accessed May 2011.

Ram Power, 2010. "Ram Power Announces Successful Well at Orita (Imperial Valley)." Marketwire. <http://www.marketwire.com/press-release/Ram-Power-Announces-SuccessfulWell-at-Orita-Imperial-Valley-TSX-RPG-1372688.htm.

Schultze, L.E., Bauer, D.J., 1982". Operation of a Mineral Recovery Unit on Brine From the Salton Sea Known Geothermal Resource Area." U.S. Bureau of Mines Report of Investigations 8680.

Tahil, W., 2007. "The Trouble with Lithium: Implications for Future PHEV Production for Lithium Demand." Meridian International Research. http://www.meridian Lithium Problem 2.pdf.

Anderson, E. R., 2011. "Shocking Future Battering the Lithium Industry through 2020." TRU Group, 3rd Lithium Supply and Markets Conference, Toronto, Ontario, Canada (January 19-21, 2011).

United States Geological Survey (USGS), 2009. "2009 Minerals Yearbook: Lithium [Advance Release]." U.S. Geological Survey, Washington, DC, USA, 11pp.

United States Geological Survey (USGS), 2011. "Mineral Commodity Summaries 2011." U.S. Geological Survey, Washington, DC, USA, pp. 94-95.

Wilburn, D.R., 2008. "Material Use in the United States—Selected Case Studies for Cadmium, Cobalt, Lithium, and Nickel in Rechargeable Batteries." U.S. Geological Survey, Washington, DC, USA. http://pubs. usgs.gov/sir/2008/5141/.

Clutter, T.J., 2000. "Mining Economic Benefits from Geothermal Brine." GHC Bulletin, Klamath Falls, OR, USA. http://geoheat.oit.edu/bulletin/ bull21-2/art1.pdf.

End Note

[1] Salton Sea and Brawley resources geographically lie within the broader Imperial Valley resource area and are commonly interchanged in the literature. For the sake of this paper, all three will be included within the broader term "Imperial Valley resource area."

[2] The USGS differentiates the terms reserve and resource on the following basis (USGS, 2011):

- Reserve: That part of the resource (sic) which could be economically extracted or produced at the time of determination. The term reserves need not signify that extraction facilities are in place and operative. Reserves include only recoverable materials; thus, terms such as "extractable reserves" and "recoverable reserves" are redundant and are not a part of this classification system.

- Resource: A concentration of naturally occurring solid, liquid, or gaseous material in or on the Earth's crust in such form and amount that economic extraction of a commodity from the concentration is currently or potentially feasible.

INDEX